Building Contractor

Start and run a money-making business

R. Dodge Woodson

TAB Books
Division of McGraw-Hill, Inc.
Blue Ridge Summit, PA 17294-0850

FIRST EDITION
FIRST PRINTING

Published by TAB Books.
TAB Books is a division of McGraw-Hill, Inc.

Library of Congress Cataloging-in-Publication Data

Woodson, R. Dodge (Roger Dodge), 1955-
Building contractor / by R. Dodge Woodson.
p. cm.
Includes index.
ISBN 0-8306-4321-4 (pbk.)
1. Construction industry—Management. 2. Contractors.
3. Construction industry—Subcontracting. I. Title.
TH438.W66 1993
690'.068—dc20 93-18549
CIP

Editorial team: Kimberly Tabor, Acquisitions Editor
Joanne Slike, Executive Editor
Barbara M. Minich, Book Editor
Joann Woy, Indexer
Production team: Katherine G. Brown, Director
Patsy D. Harne, Layout
Rhonda E. Baker, Typesetting
Jan Fisher, Typesetting
Olive A. Harmon, Typesetting
Lisa M. Mellott, Typesetting
Brenda M. Plasterer, Typesetting
Linda L. King, Proofreading
N. Nadine McFarland, Quality Control
Design team: Jaclyn J. Boone, Designer
Brian Allison, Associate Designer
Cover photograph: Brent Blair, Harrisburg, Pa.
Cover design: Stickles Associates, Allentown, Pa.

TAB1

Contents

This book is dedicated to
my wonderful daughter, Afton,
and my lovely wife, Kimberley.

Acknowledgments

I would like to acknowledge my parents, Maralou and Woody, for their years of help and support. Mom and Dad, thanks.

I thank the following government agencies for their permission to use the illustrative materials:

- U.S. Department of Agriculture
- U.S. Department of the Army
- U.S. Department of the Treasury—Internal Revenue Service

Acknowledgment and thanks are extended to Kimberly Tabor, editor-in-chief, and all of the people that have participated in the production of this book.

Introduction

Being successful as a contractor can be difficult for the most experienced business owner. For individuals just starting out, or with limited experience in business matters, success might seem unreachable. Survival can depend on experience—experience that many contractors don't have. This book is written to change all of that.

Written by a seasoned professional, this book will show you how to survive and succeed as a contractor. Regardless of your experience level, you will be able to comprehend and use this invaluable handbook.

The author, R. Dodge Woodson, is an 18-year veteran of the contracting business. Dodge is licensed as a master plumber, general contractor, consultant, and real estate broker. He has owned and operated many types of businesses. His contracting ventures have included plumbing, remodeling, and home building.

Dodge has run townhouse projects with as many as 365 units. He has been the general contractor for as many as 60 single-family homes per year. Remodeling and plumbing have become second nature to this experienced contractor. As a business consultant, Dodge specializes in helping troubled contracting companies.

After reading this book, you will probably dog-ear the pages, going back over issues you didn't even know applied to you. Between these covers lie the words of experience, both from the perspective of a contractor and a business consultant.

For less than the cost of a half-hour consultation, you will gain knowledge that could take you years to learn from your mistakes. Best of all, you don't have to make those mistakes to learn your lessons. The author is going to tell you, in easy-to-read, step-by-step instructions, how to make the most of your business.

You will learn why you need to set goals and how to set them. Chapter 1 will show you the benefits of using goals to get what you want, when you want it.

Chapter 2 puts you on the path to success. This chapter will teach you how to make a viable business plan and how to deal with recessions in the economy. You will be shown the pros and cons of corporations, sole proprietorships, and partnerships. Did you know that by making your business a standard corporation, your income may be taxed twice? Were you aware that a Subchapter S-Corporation can remove this double taxation? Well, after reading chapter 2, you will know the ins and outs of the various types of business structures.

Taxes are a turnoff for most people, but they are a fact of life. Chapter 3 covers taxes, accounting, legal considerations, and the paperwork necessary to keep your business running smoothly and profitably.

Organization is in the forefront of all successful business operations, and chapter 4 will show you how to get organized and stay organized.

Time is money, and chapter 5 shows you how to make more money by controlling your wasted time.

If your business is less than three years old, you will enjoy chapter 6. This chapter shows you how and why the first three years of your business are the hardest. But that's not all, the chapter goes on to tell you how to take the sting out of those early years. New business owners often have trouble establishing credit. Chapter 6 shows you the importance of a good credit rating and how to establish one. As a bonus, you are given seven special techniques that assure you success in your search for credit.

Managing your money for the best results in your business is as important as making money, as chapter 7 will show you.

Chapter 8 deals with the considerations you must face in choosing your office location and space.

If you are considering computerizing your business, but don't know where to begin, chapter 9 can help you. You'll see what computers can do to improve your business. Complicated computer facts are broken down into reader-friendly words and terms.

Chapter 10 concentrates on equipment, vehicles, and inventory. You will be shown the advantages of leasing vs. buying. Techniques for stocking your truck to reach maximum efficiency will be revealed. You will gain a vast amount of knowledge on equipment, vehicles, and inventory.

Chapter 11 tells you about working with subcontractors, suppliers, code officers, and materials.

Do you know how to pick the right price for your services and materials? What percentage of markup will you need to add to your job costs to be profitable? Chapter 12 answers the questions you have about pricing your goods and labor.

Schedules, budgets, and job costing are all dealt with in chapter 13. You will learn how to forecast and adjust your schedules and budgets. Examples will be given to show you how valuable job costing can be to your business.

Without customers, your business isn't worth much. Chapter 14 will show you how to become your own public relations officer and how to keep your customers happy, without giving away the store.

Can your company obtain a performance bond for a big job? Do you know the best way to build your customer base? Chapter 15 will show you how to bid jobs and how to build your business clientele.

How the public perceives your business is half the battle of finding success. Chapter 16 will show you the importance of picking the right company name and logo. You will learn how to separate your business from your competition, and more.

Many contractors feel they are too busy doing their work in the field to worry much about marketing and advertising. Unfortunately, if you don't tend to these two elements of your business, you may not have any work to do in the field. Chapter 17 explores all aspects of marketing and advertising.

If you have, or want, employees, chapter 18 is must reading. This chapter takes you by the hand and leads you through the tangled maze of employee issues.

When contractors are faced with questions on insurance, employee benefits, and retirement plans, they often throw up their hands and head off into the field. After reading chapter 19, you will have no reason to be frustrated by these complex issues. Whether you need help choosing the right insurance or aren't sure how to provide for your retirement, you will find the answers in this chapter.

When money is tight, you may have to cut back on your expenses. However, cutting the wrong expenses can cost you more than you save. Chapter 20 will educate you in what expenses to cut and which expenses to leave alone. Additionally, you will be taught how to project the future of your company.

Although good books live long lives, sometimes the material in them becomes outdated. The author has worked diligently to avoid errors and outdated information, but you should always confirm present requirements before acting on the advice given between these pages. The advice here has worked for others, and it should work for you; however no two businesses are the same. Use the following information as a guide when you make your own decisions about business endeavors.

1

Determine your business goals

Many people start a business without goals. After spending considerable sums of start-up money, they realize the business is not what they want. These people have two basic options: they can shut down the business, or they can go on with a business that does not make them happy. If they close their business, they lose money. If they continue with the business, they are no happier than they were before they started it. Neither option is desirable.

It is not uncommon for people to put themselves into business situations that they regret. For some, the stress of owning and operating a business is too much. For others, the financial ups and downs are more than they care to deal with. Being in business for yourself is not all leisure time and big bank accounts. Self-employment requires discipline, long hours, dedication, and persistence.

WHEN YOUR JOB BECOMES YOUR BUSINESS

When your job becomes your business, your life changes. Everyone hopes the change will be for the better, but that is not always the case. Owning and running your own business is not the same as going to your old job. You don't have a company supervisor to answer to, but you still have a boss. In fact, you now have many supervisors: your customers. If you don't do your job to the satisfaction of your customers, you won't be self-employed for long.

The truth is, being in business for yourself can be much more demanding than holding down a job. For example, let's say you are a carpenter. When you work for someone else, all you have to worry about is the quality of your carpentry work and the basic responsibilities of an employee, such as show up for work on time, give a fair day's production, and so on. You go to work, do the carpentry work, and go home. Once you're home, the rest of the day, evening, and night is yours. This is not usually the case when you are in business for yourself.

As a self-employed carpenter, you will have to perform all the normal carpentry duties. Then paperwork must be done, phone calls must be returned, estimates must be made, complaints must be answered and solved, marketing strategies must be defined, and accounts receivable and payable must be collected and paid. The list of additional duties goes on and on.

As you can see, running your own business is not the same as working a regular job. When your job becomes your business, you will have many more job-related responsibilities. Time with your family will be at a premium. Weekend outings might have to be postponed while you catch up on business matters that were not completed during the week.

Opening your own business is no small undertaking. The time and financial requirements of starting a business can be overpowering. Before you jump into the deep, and sometimes turbid, water of the self-employed, give careful consideration to your goals and desires.

DECIDE WHAT YOU WANT

Although this might appear simple, many people can't say what they want from their business. As a business consultant, I talk with a variety of people and businesses. When I go in to troubleshoot a business, the first question I ask the owner is what do you want from the business? More often than not, the owner doesn't know what he wants. Generally, those who give an answer, answer with a broad and unfocused reply. To be successful and to ensure the survival of your business, you must have a business plan.

When I opened my first business, as a plumbing contractor, I wanted to be my own boss. I wanted to work my own hours and not be worried about putting in 18 years, only to be let go before retirement. My dream called for building a powerful business so that I would be self-sufficient in my old age. Well, I started the business and I was relatively successful. However, looking back, I can see countless mistakes that I made.

Since that first business I have gone on to open many new businesses. Each time I start a new venture I seem to find new faults with my procedures. It is not that my methods don't work, I just always seem to find ways to improve upon them. I wouldn't begin to tell you that I know all the answers or can tell you exactly what you need to know to make your business work. But I can give you hundreds of examples of what not to do, and I can tell you what has worked in my business endeavors and those of my clients.

You never finish refining your business techniques. Even if the business climate is stable, you can always find ways to enhance your business. The business world changes frequently, forcing changes in business procedures and policies. What worked 10 years ago might not be effective today. If you are going to start and maintain a healthy business, you must be willing to change your business techniques as necessary.

SET REALISTIC GOALS

Now, back to the question, what do you want from you business? Take some time to think about the question. Then, write down your desires and goals on a sheet of paper. You *must* write down your goals and desires. For years I refused to believe that writing down my goals and desires on paper would make a difference, but it does.

After you complete your list, check over it. Break broad categories down into more manageable sizes. For example, if you wrote down that you want to make a lot of money, define how much is "a lot." Is it $30,000, $50,000, or $100,000 a year? If you jotted down a desire to work your own hours, create your potential work schedule. Will you work 8-hour days or 10-hour days? Will you work weekends? Are your scheduled hours going to comply with the needs of customers? For example, let's say you are going to open a plumbing business.

Your anticipated work schedule shows that you will work from 7:30 A.M. to 4:30 P.M., with an hour off for lunch. How realistic is this? Well, the answer depends a great deal on the type of plumbing work you will be doing. If you are working only new-construction jobs, your schedule might work. However, if you will be running a service and repair business, your schedule will never work. If a person's toilet is overflowing just after supper, it is not likely the person will be willing to wait until the next morning to have it fixed. If you are going to be competitive in the service business, you must be available for emergency calls on a 24-hour basis.

Are you willing to crawl out of your warm bed at midnight to go and fix someone's broken water pipe? If you are, you can probably charge time-and-a-half or double-time rates. The money made from night and holiday calls is very good, but if you are unwilling to take these calls, consider another form of business.

The point is this: look over your list and rate each goal and desire. You want to make a lot of money. You can do that by working overtime calls, but are you willing to sacrifice the time with your family? Go through your list with this type of question-and-answer procedure. Evaluate your goals and your willingness to achieve them. This is the first step toward starting or improving your business.

FOCUS ON THE FUTURE

Where do you want your business to be in five years? A key step toward securing a good future for your business is to develop specific goals and plans. As you create your five-year plan, ask yourself the following:

- How big do you want your business to become?
- How many employees do you want?
- Are you willing to diversify your business?

How big do you want your business to become?

Do you want a fleet of trucks and an army of employees? If you answer yes to this question, you must ask yourself more questions. Are you willing to pay the high overhead expenses that go hand-in-hand with a large group of employees? Will you need to take human resource classes to understand the steps and procedures needed to manage your employees? Will you have the knowledge to oversee accounting procedures, safety requirements, and insurance needs?

You see, in business almost every answer raises new questions. As a business owner, you must be prepared to answer all the questions. It is all but impossible for an individual to have the experience and knowledge needed to answer so many diverse questions correctly.

For example, let's say you are about to hire your first employee. Do you know what questions you can ask without violating the employee's rights? Are you aware of the laws pertaining to discrimination and labor relations? The chances are good that you don't, so what do you do? Consult with professionals in the field of expertise pertinent to your questions as you build your knowledge in these areas.

Are you willing to diversify?

As a heating contractor you might find it hard to keep money coming in during the summer months. Should you consider expanding your business to include air-conditioning work? The first consideration should be how much you know about air-conditioning. If you are capable of doing air-conditioning work yourself, adding the service to your business should increase your income. However, suppose you don't know much about air-conditioning. Should you hire a cooling mechanic to run this side of your business?

The beneficial side of adding a cooling mechanic will be increased revenue for your business. However, just because your gross sales go up doesn't mean your net profits will escalate. What will you do with the cooling mechanic during the cold months? How will you know if your employee is doing a good job? After all, you don't know much about air-conditioning. Suppose your cooling mechanic quits on short notice; what would you tell your customers? If angry customers call in July, wanting to know when their air-conditioning job will be finished, your life will not be pleasant.

I think it is healthy for businesses to diversify, but I believe the conditions must be right. In my opinion, you should not hire people to do a job about which you have no knowledge. For example, as a plumbing contractor I could hire a master electrician and expand my business base. It would be appealing to general contractors to be able to deal with one subcontractor for all their plumbing, heating, cooling, and electrical needs. So, why don't I do it? I haven't hired an electrician because I have limited knowledge of electrical wiring.

If my plumbers make a mistake or leave me out on a limb, I have the ability to work myself out of the jam. If the same happened with an electrician, I would be helpless. Without a master electrician, I would not be able to complete work that was started or repair work already installed. I see the potential dangers of expanding with electrical services to be too great for the rate of return. However, I am a capable builder, so I have no reservations about expanding as a remodeling contractor or home builder. You can apply this type of logic to any business venture. You can even use it when you evaluate whether or not going into business is the right move for you.

USE THE RIGHT MEASURING STICKS

When you are planning the destiny of your business, you must know what measuring sticks to use:

- gross sales
- number of employees
- net profits
- tangible assets,or
- some other means of comparison

Gross income is one of the most common business measurements. However, gross income can be very deceiving. Theoretically, a higher gross income should translate into a higher net income, but it doesn't always work that way. Having a fleet of new trucks might impress people and create a successful public opinion, but it also might cause your business to fail.

The best measurement of your business is the net profit. Having 10 plumbers in the field might allow you to deposit large checks in the bank, but after your expenses, how much will be left? It has been my experience, and the experience of my clients, that you must base your growth plans on net income, not gross income. Determine how much money you want to make. Then create a business plan that allows you to reach your goal.

DEVISE A BUSINESS PLAN

What is a business plan? A business plan is simply a blueprint of your business. It diagrams where you are, where you want to go, and how you are going to get there. A business plan should include as much about your business as possible. The more you put into your business plan, the better you will be able to track your success.

In my early years I never had a viable business plan. My plan was to work as hard as necessary, to make as much money as possible, and to do what I wanted to do, within reason. As time and money passed, I learned the value of a solid business plan. I can tell you, from my extensive experience, that a good business plan is instrumental to your success. A business plan does not have to be complicated, only complete.

How do you devise a business plan? Reading is one way to gain the knowledge you will need to build a business plan. Books may not be cheap, but they are generally a good value. Hiring a professional consultant is a quick way to have your questions answered. Consultants, accountants, and attorneys may seem expensive, but if they are good, they are well worth the expense. A combination of reading, seminars, and consultations is probably the quickest way to hit the fast track.

DECIDE ON THE TYPE OF WORK YOU WANT

Do you want residential or commercial business? I have always preferred residential work, but many people enjoy commercial work. This question should be easy for you to answer. You know the type of work you prefer. However, it may be worthwhile to note some of the differences between the two types of work.

Residential work

Residential work is often done on a relatively small scale. By that I mean residential work usually involves only one house. You could, however, take on a residential project that includes several hundred townhouses or condos.

Residential work offers many advantages. It is usually a more relaxed atmosphere than commercial work. The size of the jobs are generally small and can be completed in a short time. This helps contractors and subcontractors with their cash-flow needs. The amount of credit needed to finance a residential job is less than that needed for a commercial project. Work crews tend to be smaller on residential jobs, often involving only one person. This makes residential work ideal for the tradesperson just getting started in a new business.

Commercial work

Commercial work is usually more structured than residential work. Blueprints for commercial jobs are extensive and contain many details and diagrams that are not generally found on residential jobs. On the right jobs, safety might be better with commercial jobs than residential jobs, but I have always felt safe on residential jobs.

Because commercial jobs are larger than most residential jobs, the potential profit from each job is much higher. This means you can win less bids and make more money. However, big jobs often require bonding, which can be difficult for new companies to obtain. License requirements may also be stiffer for commercial jobs. The best advice I can give on choosing between commercial and residential work is to choose the field with which you are the most comfortable.

DEFINE YOUR CUSTOMER BASE

In the early days of your business, any paying customer will be welcome. However, it is important for you to determine the type of clientele with which you wish to work. The steps you take in the early months of your business influence the character of your business for a long time to come.

During the initial start-up of a business it is easy to justify taking any job that comes along. The same may be true during poor economic times. While this type of approach may be necessary, you should never lose sight of your business goals. For example, if you have chosen to specialize in new construction, make every effort to concentrate in this field of work. When you are forced into remodeling or repair work, do it to pay the bills, but continue to pursue new construction. If you bounce back and forth between different types of work, it will be more difficult to build a strong customer base and to streamline your business. Let me give you an example from my past.

When I opened my plumbing business, it was my only source of income. I wanted to be known as a remodeling plumber. After research, I had determined I could make more money doing high-scale remodeling than I could in any other field of plumbing. I reasoned that remodeling was more stable than new construction work and required less running around and lost time than service work. So I had a plan. I would become known as the best remodeling plumber in town.

As my business developed, it was tough finding enough remodeling work to make ends meet. I took on some new-construction plumbing, cleaned drains, and repaired existing plumbing. I was tempted to get greedy and try to do it all. But I asked myself, will this approach work? I decided it wouldn't because of the nature of the different types of work.

With new-construction, bid prices were competitive. To win the job and make money, I had to work fast and eliminate lost time. If I was plumbing a new house and my beeper went off, I had to pick up my tools and stock and leave the job to call the answering service. Then I had to call the customer. Then I had to either respond to the call or try to put the customer off until I left the new-construction work.

Instead, I set my sights on remodeling and put all my effort into getting remodeling jobs. In a matter of months, I was busy and my customers were happy. My net income rose because I had eliminated wasted time. In time, I added more plumbers and built a solid service and repair division. Then I added more plumbers and took on more new-construction work. But you see, you must carefully structure your business plan or it will get away from you and cause you to work harder, while making less money.

When you consider your desired customer base and work type, do it judiciously. If you live in a small town, you may not be able to specialize in framing, roofing, or remodeling. You may have to do a little of everything to stay busy. But never lose your perspective. Pursue the type of work and customers you desire.

WRITE YOUR JOB DESCRIPTION

For most new entrepreneurs, it is necessary to play all positions. However, just like setting a goal for the type of work your business will do, you should establish a goal for the type of work you want to do. Do you want to work in the field or in the office? Will you trust important elements of your business to employees? Will you want to do it all yourself?

Delegating duties is difficult for many first-time business owners. Most people fall into one of two traps. The first group believes that they must do everything themselves. This group hesitates to delegate duties, and even after assigning the task to competent employees, keep their fingers in the work. The second group believes everything can be delegated and passes responsibilities on to people that are not in a position to make proper judgments. Successful business owners fit somewhere between these two extremes.

It is counterproductive to hire employees if you are not going to allow them to do their jobs. You should supervise and inspect the employees' work, but you should not look over their shoulders every five minutes. If you did a good job screening and hiring your employees, they should be capable of working with limited supervision. When you spend time hovering over employees, you neglect many of your management and ownership duties.

If you decide to do everything yourself, you must recognize the fact that a time will come when you have to turn away business. One person can only do so much. It is better to politely refuse work than it is to take on too much work and not get it done.

Will you be content to stay in an office? If your nature tends to keep you outside and doing physical work, being office-based can be a struggle. It has its advantages, but office work can be a real drag to the person accustomed to being out and about.

If you don't want to work in the office, you must make arrangements so you have your freedom. Answering services and machines afford some relief. A receptionist is another way to keep the office staffed while you are out, but this is an expensive option. If you will be in the field, a pager and cellular telephone may be your best choice for keeping in touch with your customers. It will be up to you to devise the most efficient way to stay on the job and out of the office, but don't overlook this question when you plan your business.

Now for the reverse situation, suppose you want to be in the office. Who will be in the field? The solution to this problem is often not easy for a new business owner. If you hire employees to do the field work, you will need to have enough work to keep them busy. Getting steady work is rarely simple. For a new business it can be nearly impossible. I always run into the same problem in this situation. I either have too many mechanics and not enough work, or too much work and not enough mechanics.

Most new businesses are run by the owners. For tradespeople, it is common for them to do field work during the day and office work at night. While this is usually mandatory, it doesn't have to stay this way forever. Decide where you want to be, in the field or in the office, and work up a plan that will enable you to meet your goal.

2

Define your business structure

Corporations, sole proprietorships, and partnerships are all common forms of business structure. Each form of business structure has its own advantages and disadvantages. This chapter is going to help you explore the different types of business structure and decide which one best suits your needs.

You may find that incorporating your business will provide you with some security from personal liability. You might discover that if you incorporate, your tax consequences increase. The lure of partnerships may not be so attractive after reading about what can happen to you. By the time you finish this chapter, you should know what type of business structure is best for you.

EVALUATE YOUR OPTIONS

Choosing your style of business structure is not a task to take lightly. There are a multitude of considerations to evaluate before you decide on a business structure for your venture. Do you know the difference between a standard corporation and a Subchapter S-Corporation? Are you aware of the tax advantages to running a sole proprietorship? Were you aware that having a partner can ruin your credit rating? If you answered no to these questions, you will be especially glad you read this chapter.

Before we get started, let me say that I am not an expert on tax and legal matters. The information in this chapter is based on experience and research. As with all the information in this book, you should verify the validity of the information before using it. Laws change and different jurisdictions have different rules. Now, with that out of the way, let's dig into the various types of business structures.

Corporations

A corporation is a legal entity. To be legal, a corporation must be registered with and approved by the secretary of the state in which it is doing business. A corporation can live on in perpetuity. Corporations may operate under general management, may provide limited liability, and may transfer shares of stock.

Subchapter S-Corporations

A Subchapter S-Corporation is not like a standard corporation. Commonly called an S-Corporation, this corporate structure must meet certain requirements. For example, an S-Corporation may not have more than 35 stockholders. An S-Corporation is one that chooses not to be taxed in the same way as a regular corporation.

Additional requirements for S-Corporations include having stockholders provide personal tax returns. The stockholders must show their share of capital gains and ordinary income. Tax preference also must be provided personally by the stockholders. The corporation files a tax return, but does not pay income tax. This type of corporation allows the stockholders to avoid double taxation. However, Subchapter S-Corporations may not receive more than 20 percent of their income from passive income. Passive income might be rent from an apartment building, book royalties, earned interest, and so on.

Partnerships

A partnership is an agreement between two or more people to do business together. In a general partnership, each partner is responsible for all of the partnership's debt. This is an important fact to know and understand. If you have a bad partner, you could be held liable for all partnership debts your partner incurs. Most partnerships don't pay income taxes, but must file a tax return. The income from the partnership is taxed through the personal tax returns of the partners.

Partnerships may be comprised of all general partners or one general partner and any number of limited partners. General partners do not have limited liability. The general partner can be held accountable for all actions of the partnership. Limited partners may have limited risk. For example, a limited partner may invest $10,000 in the partnership and limit his liability to that $10,000. Then, if the partnership loses money, goes bankrupt, or has other problems, the most that the limited partner stands to lose is his investment of $10,000. The general partner, however, will be held personally responsible for all actions of the partnership.

Sole proprietorships

A sole proprietorship is a business owned by an individual. The business must file a tax return, but any income tax is assessed against the individual's tax return. Sole proprietors are exposed to full liability for their business actions.

Now that you know about the different types of business structure, what type would you choose? The truth is, I haven't given you enough information to make a wise decision. Before you make a firm decision, you owe it to yourself to read and digest the remainder of this chapter. It will be interesting to see if your choice after reading the entire chapter is the same as it is now.

SHOULD YOU INCORPORATE?

Corporate advantages are abundant for large businesses, but a standard corporation may not be a wise decision for a small business. Depending on the size and operational aspects of your business, a corporation might provide protection from personal lawsuits and financial problems. Since corporations can issue stock, there is the possible advantage of generating cash from stockholders. Large sums of money can be generated when a corporation goes public, but this is not usually a feasible option for a small business.

While small businesses might suffer from double taxation as a standard corporation, they can benefit from some tax advantages. If you incorporate your business, some expenses that are not deductible as a noncorporate entity become deductible. Insurance benefits are an example of this type of deductible advantage.

In many ways, there are more disadvantages to incorporation than advantages for the small business. The first disadvantage is the cost of incorporating. There are filing fees that must be paid, and most people use lawyers to set up corporations. The cost of establishing a corporate entity can range from less than $200 to upwards of $1,000. This can be a lot of money for a fledgling business.

There is more to having a corporation than setting it up and forgetting about it. To keep the advantages of a corporation effective, you must maintain certain criteria. You will have to have a registered agent. Many people use their attorney as a registered agent. If you use an attorney in this position, you will be spending extra money. Written minutes must be recorded at these meetings. Board meetings must be held and the corporate book must be maintained. A board of directors and corporate officers must be established. Written minutes must be recorded at these meetings. Annual reports are another responsibility of corporations. Again, many people have an attorney tend to much of this work.

The fees for maintaining an effective corporation can add up. If the corporate rules are broken, the corporation loses much of its protection potential. If an aggressor can pierce the veil of your corporation, you will have personal liability. Small business owners that incorporate their business with a standard corporation will face double taxation. These owners will pay personal income tax and will also pay corporate taxes. This extra taxation is a serious burden for most small businesses.

The protection gained from incorporating may not be as good as you think. Let's explore some of the misconceptions about corporate protections.

Lawsuit protection

It is almost impossible to protect yourself entirely from lawsuits, but a corporation can help. If your business is a corporation, your liability may be limited to the corporate assets, but don't count on it.

Assume that your business is a corporation. You are an electrician. You go out and install new wiring in a remodeling job. Being a one-employee corporation, you do all the work yourself. Later that night someone is injured by an electrical shock from your new work. Your customer decides to sue. Can the customer sue the corporation and go after its assets? Yes, the customer can sue the corporation. Can the customer sue you, as the primary stockholder of the corporation? I suppose anyone can sue anyone else for any reason, but it would be unlikely that the customer would get far in a lawsuit against you as a stockholder. But don't relax, you're not safe. Since you did the work yourself, the customer can sue you individually for the mishap. This is a fact that most business owners don't realize.

If you had sent an employee or a subcontractor out to do the work, your risk of being sued for your personal assets, other than your corporation, would be minimal. So if you don't do your own field work, a corporation can help protect your personal assets.

Personal financial protection

People feel they can limit their financial liability by incorporating, and to some extent this is true. If a corporation gets in financial trouble, it can declare bankruptcy, without the stockholders losing their personal assets. However, most lenders and businesses that extend credit to corporations require someone to sign personally for the debt. If you personally endorse a corporate loan, you will be responsible for the loan even if the corporation goes belly up. Before you spend needed money on false protection, talk to experts about how incorporating will affect you and your business.

CONSIDER A SUBCHAPTER S-CORPORATION

One of the biggest advantages of a Subchapter S-Corporation over a standard corporation is the elimination of the double taxation. Stockholders of Subchapter S-Corporations pay only personal taxes on the money they and the corporation earn. S-Corporations offer the other advantages you would receive with a standard corporation.

The cost of setting up and maintaining the corporations are still there. You will not be able to use an S-Corporation once you have more than 35 stockholders. If more than 20 percent of your corporate income will be passive income, you may not use a Subchapter S-Corporation.

BE WARY OF PARTNERSHIPS

The cost of establishing a partnership is less than incorporating, but I don't see the need for partnerships in small business operations. Partnerships have their place in business ventures, but I don't like them for most types of business. If you are a real estate investor, partnerships can work. But why gamble with a partnership for a service business. Set up a corporation or investment agreements, but avoid partnerships.

Partnerships should be treated seriously. If you are a general partner, you can be held accountable for the business actions of your partner or partners. This fact alone is enough to make partnerships a questionable option. There are very few advantages to partnerships, but the potential problems are numerous.

If you decide you want to set up a partnership, consult an attorney. The money you spend in legal fees to form the partnership will be well spent. It is much better to invest in a lawyer before you form the partnership, than to pay legal fees to resolve partnership problems later.

A SOLE PROPRIETORSHIP MAY BE BEST

A sole proprietorship is simple and inexpensive to establish. You are your business, so there are no complicated corporate records to keep. You can get tax advantages as a sole proprietor, as you can with other business structures. You are the boss, there is no partner to argue with over business decisions. There are no stockholders. Tax filing is relatively simple, and you don't have to share your profits with others.

As a sole proprietor, you will have to sign personally for all your business credit. If your business is sued, you will be sued. There may be some tax angles that you will miss out on as a sole proprietor. To find out, consult a professional.

The basic advice in this chapter is to seek competent, professional help before you set up any type of business. I think most of you will find a sole proprietorship or a Subchapter S-Corporation best suited to your business needs.

3

Complete the paperwork

Taxes, accounting, legal considerations, and the paperwork that goes along with them are all serious aspects of any business. These areas of business confuse and intimidate many business owners. Some business owners try to take an out-of-sight, out-of-mind attitude toward these less-than-desirable business responsibilities. However, this approach doesn't work for long.

Professional help is not inexpensive, but it is better to pay an attorney to draft good legal documents than it is to pay to have improper documents defended in court. The same can be said about accountants. A good certified public accountant (CPA) can save you money, even after the professional fees you are charged.

Using and maintaining the proper paperwork can make all aspects of your business better. But when it comes to taxes and legal issues, organized paperwork is invaluable. If you ever have to go to court, you will learn the value of well-documented notes and agreements. An Internal Revenue Service (IRS) audit will prove the importance of keeping good records.

THE LEGAL SIDE OF BUSINESS

Business law is a vast subject, much too broad to cover in a single chapter. If you are not well-versed in business law, spend some time studying the topic. You might find it necessary to go to seminars or college classes to gain the knowledge you need. Books that specialize in business law may be all the help you need. If you have specific questions, you can always confer with an attorney who concentrates on business law.

It is your responsibility as a business owner to comply with all laws. Ignorance of the law is not a suitable defense. The penalities for breaching some of these laws are extensive and may involve imprisonment and cash fines.

LICENSES

Licenses are required for most businesses. Some businesses only require a general business license. Other types of businesses require the standard business license, and additional licensing. For example, a plumbing business must have a business license and a master plumber's license. The same is true for electrical businesses. However, a painting business doesn't require a painter's license, only a business license.

Having or obtaining the proper licenses is the responsibility of the business owner. If you operate a business without the required licenses, you can get into trouble from all angles. Consumers will have grounds to be upset, and local authorities will have something to say about your failure to be licensed. The penalties for operating a business without a proper license can be steep. Check with local officials to see what licenses you must obtain.

CHOOSING AN ATTORNEY AND ACCOUNTANT

Finding a professional with the experience, knowledge, and skill that you require can take some time. These two professional fields incorporate an enormous amount of facts and requirements under two simple names. An individual attorney cannot possibly be fluent in all areas of the law. Accountants cannot be expected to know every aspect of financial law and practice. For these reasons, you must look for professionals who specialize in the type of service you require.

Finding a specialist is easy. Most professionals list their specialties in their advertising. Once you narrow the field to professionals working within the realm of your needs, you must further separate the crowd. Finding 10 attorneys that work with business law is not enough. You must look through those 10 lawyers and find the one that is best suited to work with your business.

Begin the process of elimination by making technical considerations. Make a list of your known and expected needs. Ask the selected professionals how they can help you with these needs. Inquire about their past performances and ask for client references.

Once the professionals have answered all your technical questions, ask yourself some questions. How do you feel about the individual professional? Would you be comfortable going into a tax audit with this CPA? Are you willing to bet your business on the knowledge and courtroom prowess of this attorney? Are you comfortable talking with the individual?

How you feel toward the professional as a person is important. You will very likely expose your deepest business secrets to your accountant and attorney. If you are not comfortable with the professional, you will not get the most out of the business relationship.

Another key factor to assess before you choose a professional is the ease with which you understand them. The subject matter you will be discussing is complex and possibly foreign to you. You need professionals who can decipher the cryptic information that confuses you and present it in an easy-to-understand manner.

TAXES ARE A FACT

Tax laws are complex and can be confusing. For the average business owner, many tax advantages go unnoticed. These business owners know their businesses, but they don't know all they should about income taxes. Unless you are a tax expert, consult with someone who is.

If you wait until a month before tax time to meet with a tax specialist, there may not be much the expert can do for you, short of filing your return. However, if you consult with tax experts early in the year, you can manage your business to minimize the tax bite. Early consultations can result in significant savings for you and your business.

Tax manipulation is an art, and CPAs are the artists. When you meet with these professionals, they can find numerous ways for you to save on your taxes. You might be told to lease vehicles, instead of buying them. You may be shown how to keep a mileage log for your vehicle to maximize your tax deductions. A CPA might recommend a different type of structure for your business. For example, if you are operating as a standard corporation, the accountant might suggest that you switch to a SubChapter S-Corporation to avoid double taxation.

If you work from home, your tax expert can show you how to deduct the area of your home that is used solely for business. A tax specialist can show you how to defer your tax payments to a time when your tax rate may be lower. Investment strategies can be planned to make the most of your investment dollars. You can coordinate a viable retirement plan by talking with an authority on tax issues.

Perhaps one of the most important things a tax specialist can teach you is the difference between legitimate and illegitimate deductions. I see numerous businesses where the business owners are required to pay back taxes. Most of these business owners had no idea they owed taxes until they got a notice to pay them. These people find that deductions they thought were legitimate were not allowed. Catching up on taxes that were due last year or years before can be a major burden. This type of expenditure is never planned for, and can drive a business into financial hardship.

The best way to avoid unexpected tax bills is to make sure your taxes are filed and paid properly. The most effective way to ensure that your taxes are done properly is to hire a professional to do them for you. If you resent paying someone to do a job that you think you can do, you are not alone. But you might find that by paying a little now, you will save a lot later. The appendix contains examples of tax forms that may be used by your business.

SURVIVING AN IRS AUDIT

If you are concerned about surviving an IRS audit, don't worry. I did it and you can do it. For years, an audit was one of my greatest fears. Even though I knew, or at least thought, that my tax filings were in order, I worried about the day I would be audited. It was a lot like dreading the semiannual trip to the dentist. Then one day it happened. I was notified that my tax records were going to be audited. My worst nightmare became a reality.

Before the audit, I scrambled to gather old records and went over my tax returns for anything that might have been in error. I couldn't find any obvious problems. I went to my CPA and had him go over the information I would present in the audit. We couldn't find any blazing red flags that called attention to my tax return.

When I called the individual that was to perform my IRS audit, I was told that my return had been chosen at random. The person went on to say that there probably was nothing wrong with my return, that a percentage of returns are picked each year for audit. Gaining this information made me feel a little better, but not a lot.

My CPA represented me on the day of the audit. I was not required to attend the meeting. Even though I wasn't at the meeting, my mental state was miserable that day. When my CPA called, he told me the meeting had gone well, but that I had to provide further documentation for some of my deductions. These deductions primarily were travel expenses and books.

After digging through my records, I found most of the needed documentation. However, there were some receipts that I couldn't document. For example, I had written on receipts that they were for book purchases, but I hadn't listed the title of the book. The IRS wanted to know what books I had purchased. I couldn't remember what the titles were, so I put notes on the receipts to explain that I could not document the book titles.

The travel expenses were easier to document, most of them had been paid for with credit cards. By finding the old receipts I was able to remember where I had gone and why. After finding as much documentation as I could, the package was given to my CPA.

Another meeting between the auditor and my accountant took place. I was expecting the worst, and thought I would have to pay back taxes on the items I couldn't identify properly. But to my surprise, I was given a basically clean bill of health. The auditor accepted my I-can't-remember receipts. Since most of my receipts were documented and I hadn't written fictitious names on the receipts, my honesty prevailed.

I always thought an audit would be horrible. However, I didn't have to attend the audit personally, my CPA did a great job, and I didn't have to pay any serious tax penalties. My audit was over, and the pain wasn't too great. I can't say that all audits are this easy, but mine was.

Going through that audit convinced me of what I had believed for years: accurate records are the key to staying out of tax trouble. If my records had been misplaced or substantially incomplete, I could have been in a serious bind. However, my good business principles enabled me to survive the audit with relative ease.

If you are afraid of being audited, begin keeping detailed records. Before taking a questionable deduction, check with a tax expert to confirm the viability of the deduction. Don't cheat on your taxes; not only is it wrong, you never know when you will be caught. Honesty is the best policy.

By keeping cash receipts for all your cash purchases, you can take advantage of all your tax deductions. Receipts will serve as documentation for your deductions. If the receipt does not state clearly what the purchase was, write the details on the receipt. If the item was for a particular job, write the job name on the receipt. If you are audited, your documented cash receipts may mean the difference between an easy audit and having to pay back taxes.

DOCUMENT YOUR BUSINESS ACTIVITY

Good documentation is essential to running a successful business. The documentation may be used to track sales, to keep up with changes in the market, to forecast the future, to defend yourself in court, or to substantiate your tax filing, just to name a few. Proper documentation can save you a tremendous amount of time and frustration.

Consider this simple example: you agree to do a job for a customer. You understand the customer to say she wants the deck on her house replaced. The customer actually only wants the deck repaired. The customer leaves for her job and you go to work. When the customer returns, she finds the deck demolished and a new one being built. You can imagine the argument that is likely to ensue.

The customer refuses to pay for the new deck, but insists that you build it to replace the one you destroyed. The path of least resistance here would be to build a new deck for the lady. However, you might decide to go to court. In court, if you don't have documentation to prove otherwise, the customer's word is as good as yours. Who will win this court battle? Your guess is as good as mine, but you don't want to leave your business in the hands of a judge or jury.

If you had a written contract, you wouldn't be having this dispute. When it comes to doing business, you must document your activity. A contract would stand up as evidence in court. The same can be said for other forms of written documentation.

There are many good reasons for and ways to document your business activity. A carbon-copy phone-message book is a simple, yet effective way to log all of your phone activity. Written contracts are instruments that document your job duties and payment arrangements. Change orders and addendums are the best way to document your actions when situations deviate from the original agreement. Letters can be used to confirm phone conversations and create a paper record that avoids confusion. Tape recorders can be used to record your daily duties. The ways to and reasons for documenting your business activity are many.

While you might abhor doing paperwork, you must recognize it as a necessary part of doing business. The use of computers has helped reduce the amount of paper used, and computers have made the task of doing paperwork easier, but even computers cannot eliminate the need for accurate paperwork. To understand the need for so much clerical work, let's take a closer look at the various needs of business owners.

Contracts

Contracts are a way of life for contractors (FIGS. 3-1 and 3-2). The contracts that a contractor enters into are the lifeblood of the business. There are two types of contracts—oral and written. Oral contracts are legal, but they es-

Letter of Engagement

Client ____

Street ____

City/State/Zip ____

Work phone ______________________ Home phone ______________________

Services requested____

____

____

____

____

____

Fee for services described above $______________________________

Payment to be made as follows:

____

____

____

By signing this letter of engagement, you indicate your understanding that this engagement letter constitutes a contractual agreement between us for the services set forth. This engagement does not include any services not specifically stated in this letter. Additional services, which you may request, will be subject to separate arrangements, to be set forth in writing.

A representative of ______________________ has advised us that we should seek legal counsel prior to using information or material received from ______________________.

We the undersigned hereby release ______________________, its employees, officers, shareholders, and representatives from any liability. We understand that we shall have no rights, claims, or recourse and waive any claims or rights we may have against ______________________, its employees, officers, shareholders, and representatives. We further understand that we will pay all costs of collection of any amount due hereunder including reasonable attorney's fees.

______________________ ______________________

Client Date Client Date

____

Company Representative Date

3-1 Letter of engagement.

Renaissance Remodeling
357 Paris Lane
Wilton, Ohio 55555
(102) 555-5555

REMODELING CONTRACT

This agreement, made this _____th day of ____________________, 19____, shall set forth the whole agreement, in its entirety, between Contractor and Customer.

Contractor: Renaissance Remodeling, referred to herein as Contractor.

Customer: ____, referred to herein as Customer.

Job name: ____

Job location: ____

The Customer and Contractor agree to the following:

Scope of Work

Contractor shall perform all work as described below and provide all material to complete the work described below: All work is to be completed by Contractor in accordance with the attached plans and specifications. All material is to be supplied by Contractor in accordance with attached plans and specifications. Said attached plans and specifications have been acknowledged and signed by Contractor and Customer.

A brief outline of the work is as follows, and all work referenced in the attached plans and specifications will be completed to the Customer's reasonable satisfaction. The following is only a basic outline of the overall work to be performed:

____

____

____

____

____

____

____

____

(Page 1 of 3 initials______________)

3-2a Remodeling contract (page 1).

Commencement and Completion Schedule

The work described above shall be started within three days of verbal notice from Customer; the projected start date is ____________. The Contractor shall complete the above work in a professional and expedient manner, by no later than _____ days from the start date. Time is of the essence regarding this contract. No extension of time will be valid, without the Customer's written consent. If Contractor does not complete the work in the time allowed, and if the lack of completion is not caused by the Customer, the Contractor will be charged _______, per day, for every day work is not finished beyond the completion date. This charge will be deducted from any payments due to the Contractor for work performed.

Contract Sum

The Customer shall pay the Contractor for the performance of completed work, subject to additions and deductions, as authorized by this agreement or attached addendum. The contract sum is ____, ($______________).

Progress Payments

The Customer shall pay the Contractor installments as detailed below, once an acceptable insurance certificate has been filed by the Contractor, with the Customer:

Customer will pay Contractor a deposit of ____,

($____________), when work is started.

Customer will pay ____,

($___________), when all rough-in work is complete.

Customer will pay ____,

($_______________) when work is _________ percent complete.

Customer will pay ____,

($____________) when all work is complete and accepted.

All payments are subject to a site inspection and approval of work by the Customer. Before final payment, the Contractor, if required, shall submit satisfactory evidence to the Customer, that all expenses related to this work have been paid and no lien risk exists on the subject property.

Working Conditions

Working hours will be ___ A.M. through ___ P.M., Monday through Friday. Contractor is required to clean work debris from the job site on a daily basis and to leave the site in a clean and neat condition. Contractor shall be responsible for removal and disposal of all debris related to their job description.

(Page 2 of 3 initials____________)

3-2b Remodeling contract (page 2).

Contract Assignment

Contractor shall not assign this contract or further subcontract the whole of this subcontract without the written consent of the Customer.

Laws, Permits, Fees, and Notices

Contractor is responsible for all required laws, permits, fees, or notices required to perform the work stated herein.

Work of Others

Contractor shall be responsible for any damage caused to existing conditions. This shall include work performed on the project by other contractors. If the Contractor damages existing conditions or work performed by other contractors, said Contractor shall be responsible for the repair of said damages. These repairs may be made by the Contractor responsible for the damages or another contractor, at the sole discretion of Customer.

The damaging Contractor shall have the opportunity to quote a price for the repairs. The Customer is under no obligation to engage the damaging Contractor to make the repairs. If a different contractor repairs the damage, the Contractor causing the damage may be back-charged for the cost of the repairs. These charges may be deducted from any monies owed to the damaging Contractor.

If no money is owed to the damaging Contractor, said Contractor shall pay the invoiced amount within _____ business days. If prompt payment is not made, the Customer may exercise all legal means to collect the requested monies. The damaging Contractor shall have no rights to lien the Customer's property for money retained to cover the repair of damages caused by the Contractor. The Customer may have the repairs made to his satisfaction.

Warranty

Contractor warrants to the Customer all work and materials, for one year from the final day of work performed.

Indemnification

To the fullest extent allowed by law, the Contractor shall indemnify and hold harmless the Customer and all of their agents and employees from and against all claims, damages, losses and expenses.

This Agreement entered into on ____________________________, 19_____ shall constitute the whole agreement between Customer and Contractor.

____ ____

Customer Date Contractor Date

____

Customer Date

3-2c Remodeling contract (page 3).

sentially are unenforceable. Written contracts are the other choice, and these agreements are enforceable.

Contracts give a full description of the work that you are being engaged to perform. When written properly, contracts leave little room for misunderstandings. Working a job without running into confused confrontations with the customer will be more pleasurable and profitable.

Good contracts include the date work will be started and the hours during which you may work. If you have a customer that doesn't want anyone on the job until 10:00 A.M., you will have trouble scheduling your workers for maximum efficiency. If you know this in advance, you can adjust your price to reflect the inconvenience. However, if you take the job and don't find out about the allowable work hours until a raging customer chases your workers off the property at 8:00 A.M., you will have problems on the job.

By including in the contract a starting date for the job, both you and the customer know when to expect the work to begin. This stops the customer from calling and hounding you for a start date. It also allows you to organize your schedule and prepare for the job. Some contractors need this committed discipline to run their businesses effectively.

Contracts should state how much you will be paid, and in what increments and at what times you will receive your money. If you are to get a deposit before you start the job, the date the deposit is due and the amount of the deposit should be stipulated in the agreement. Money is a major cause of job-related problems. The more documentation you have on the financial aspects of the job, the better off you will be.

Most contracts will go on to cover a wide variety of other variables, including who is responsible for cleaning up after the job, what will happen with the debris from the job, and the length of the guarantee. Generally speaking, a written contract is the foundation of your business.

Addendums

Addendums are extensions of a contract (FIG. 3-3). Addendums are used to add language to contracts after the contracts are written. In some cases, where fill-in-the-blank contracts are used, addendums provide a means for making the contract more explicit. Addendums will come in handy to document significant changes in the job. By documenting all agreements between yourself and the customer, you eliminate confusion and many of the risks of confrontations.

Change orders

Change orders (FIGS. 3-4 and 3-5) are written agreements that are used when a change is made to a previous contract. For example, if you are contracted to paint a house white and the customer asks you to change the color to beige, use a change order. If you paint the house beige without a change order, you could be found in breach of your contract, since the contract called for the house to be painted white.

Addendum

This addendum is an integral part of the contract dated ____________________, between the Contractor, ____, and the Customer(s), ____, for the work being done on real estate commonly known as __________________________. The undersigned parties hereby agree to the following:

____

____

____

____

____

____

____

____

____

____

____

____

____

____

____

The above constitutes the only additions to the above-mentioned contract, no verbal agreements or other changes shall be valid unless made in writing and signed by all parties.

______________________________ ______________________________

Contractor Date Customer Date

____

Customer Date

3-3 Addendum.

Change Order

This change order is an integral part of the contract dated______________________, between the customer, ________________________________ , and the contractor,____________________, for the work to be performed. The job location is ____.
The following changes are the only changes to be made. These changes shall now become a part of the original contract and may not be altered again without written authorization from all parties.

Changes to be as follows:

____

____

____

____

____

____

These changes will increase/decrease the original contract amount. Payment for theses changes will be made as follows:________________________________. The amount of change in the contract price will be ___________________________ ($). The new total contract price shall be _______________________ ($).

The undersigned parties hereby agree that these are the only changes to be made to the original contract. No verbal agreements will be valid. No further alterations will be allowed without additional written authorization, signed by all parties. This change order constitutes the entire agreement between the parties to alter the original contract.

______________________________ ______________________________
Customer Contractor

______________________________ ______________________________
Date Date

____
Customer

____
Date

3-4 Change order.

Request for Substitutions

Customer name: ______________________________

Customer address: ______________________________

Customer city/state/zip: ______________________________

Customer phone number: ______________________________

Job location: ______________________________

Plans & specifications dated: ______________________________

Bid requested from: ______________________________

Type of work: ______________________________

The following items are being substituted for the items specified in the attached plans and specifications:

____

____

____

____

____

____

____

____

____

____

Please indicate your acceptance of these substitutions by signing below.

______________________________ ______________________________

Contractor Date Customer Date

____

Customer Date

3-5 Request for substitutions.

It might seem unlikely that the customer would sue you for doing what you were told, but it is possible. Without a written change order, you would be at the mercy of the court. If the court accepted the contract as proof that the house was to be painted white, and it probably would, you would be in trouble.

Change orders should include basic information: the name and address of the customer, the location of the job, and the date and reference number of the original contract. The body of the change order should include a description of the work being altered. For example, don't write a change order that says simply that the kitchen sink will be changed from a single-bowl sink to a double-bowl sink. Include all pertinent data on the new sink: make, model number, color, style, and whatever else is applicable.

Always document how the change order will affect the cost of the job. If the change will result in a credit to the customer, put the amount of the credit in the paperwork. If the change will be reason for you to charge more money for the job, detail the extra charges in writing. Dictate how the credits or extra charges will be accounted for and when they will be due.

Require all parties that signed the original contract to sign the change order. I have seen many contractors write up a change order and give it to the customer, without ever receiving a signed copy for their files. I have also seen change orders signed by only one of the parties of the original contract. Both of these practices are bad business. Treat change orders with the same respect you do a contract, and keep signed copies on file.

Service orders

Most companies that provide routine maintenance and repair services use service orders. Service orders are the small tickets that customers are asked to sign, generally after the work is done, to acknowledge that the work was done satisfactorily. These forms are fine, except for the fact that most customers are not asked to sign the service order until the work is completed. This practice puts the business owner at risk. The customer should be asked to sign the service order before the work is started and again when the work is completed.

The wording in your mini-contract should include payment terms, guarantees, liability restrictions, and much more. An attorney can help you design a service order that will best suit your needs.

Use service orders to document the time your employee arrives on a job and the time he leaves. This helps track your employee's productivity and gives you an edge in management.

Service orders can also help you with inventory control. Have your service technicians list all the materials they used on the service ticket. You can then maintain an accurate inventory of your rolling stock. When designed and used properly, service orders can help you better manage your business.

Liability waivers

Liability waivers are written releases of liability. These forms protect you from being accused of an act that was nearly unavoidable. Use liability waivers any time you believe a confrontation might arise from your actions. For example, assume you are a plumber. A customer asks you to replace the flush valve in a very old toilet. Being an experienced plumber, you know that when trying to loosen the nut on an old flush valve there is a risk of breaking the toilet. Because of the possibility of damaging the toilet, you have the customer sign a liability wavier. This way, if the toilet is broken accidentally, you are not held responsible for the damage.

You probably won't find many preprinted liability waivers. I suspect that in time companies will make generic waivers, much like the fill-in the-blank contract forms, but for now you will need to consult with an attorney.

When you talk with your attorney about a liability waiver, explain all aspects of your business. The attorney won't be able to give you a comprehensive liability waiver unless you are specific about what you do. While you are talking with your lawyer, ask for some boilerplate language that you can use for on-the-spot cases. Even a customized waiver form will not fulfill all potential needs. There will come a time when special circumstances will require you to draft a liability waiver on the job.

First, check to make sure you legally can draft your own waiver. Some states are specific on what legal documents individuals can prepare. If you can create your own liability waiver, obtain suitable language for the waiver from your attorney. If you are prohibited from drawing up your own document, let your attorney make a fill-in-the-blank waiver. Then you can add specific details to the waiver as required.

Written estimates

Written estimates (FIG. 3-6) reduce the risk of confusion when you give prices to customers and tell them your terms and conditions. They also present a more professional image than oral estimates. Giving quotes verbally can lead to many problems. To eliminate these problems, make all of your estimates in writing and keep copies of the estimates.

Specifications

Written specifications (FIG. 3-7) are another way to avoid confrontations with customers. With small jobs, the specifications can be included in the contract. When you are embarking on a large job, the specifications will generally be too expansive to put in the contract. In these instances, make reference in the contract to the specifications. Then attach the specification sheet to the contract and make it a part of the contract.

A complete set of written specifications is beneficial to you and your customer during the course of the job. If you take the time to document all of the job specifications, you are less likely to lose time arguing with a dissatisfied customer. This not only saves time and money, it helps keep your customers happy.

Green Tree Lawn Care
987 Willow Road
Wilson, Maine 55555
(101) 555-5555

WORK ESTIMATE

Date: ______________________________

Customer name: ____

Address: ____

Phone number: ____

Description of Work

Green Tree Lawn Care will supply all labor and material for the following work:

____

____

____

____

____

____

____

____

Payment for Work as Follows

Estimated price: ____, payable as follows

____

____

If you have any questions, please don't hesitate to call. Upon acceptance, a formal contract will be issued.

Respectfully submitted,

J. B. Williams
Owner

3-6 Work estimate.

Material Specifications

Phase	Item	Brand	Model	Color	Size
Plumbing	Lavatory	WXYA	497	White	19"×17"
Plumbing	Toilet	ABC12	21	White	12" rough
Plumbing	Shower	KYTCY	41	White	36"×36"
Electrical	Ceiling fan	SPARK	2345	Gold	30"
Electrical	Light kit	JFOR2	380	White	Standard
Flooring	Carpet	MISTY	32	Grey	14 yards

3-7 Material specifications.

Be as detailed as possible when you develop specifications for a job. Include model numbers, makes, colors, sizes, brand names, and other suitable descriptions of the labor and materials you will be providing. Once you have a good spec list, have the customer review and sign it. Without a signature of acceptance from the customer, your specs are little more than a working guide for you. Unsigned specification sheets carry little weight in a legal battle.

Customer credit applications

Credit applications should be filled out by all customers to whom you are planning to extend credit. These forms will allow you to check into the individual's past credit history. Just because a person has good references now doesn't mean payment is guaranteed, but your odds for collecting the money owed are better if your customer has a history of paying on time.

As a business owner you may subscribe to the services of a credit reporting bureau. For a small monthly fee and an inexpensive per-inquiry fee, you can pull a detailed credit history on your customers. You will, of course, need the permission of your customers to check into their credit background. Credit applications provide documentation of this permission.

Even if you don't belong to a credit bureau, you can call references given by the customer on the credit application. This type of investigation is not as good as the reports you receive from credit agencies, but it is better than nothing. Credit applications should be used for every customer who wishes to establish a credit account with you.

Inventory logs

Inventory logs (FIG. 3-8) should be used to maintain current information on your inventory needs and supply. If you take materials out of your inventory, write on the log what the items were and where they were used. At the end of the week, go over your log and adjust your inventory figures. If you need to replace inventory, you will know exactly what was used. If questions arise at tax time, you can identify where your inventory went. Something as simple as an inventory log can save you money and time.

Inventory Log

Item	*Quantity*	*ID Number*	*Checked By*	*Date*

3-8 Inventory log.

Repair vouchers

If customers can bring items to your business for repair, you should use repair vouchers (FIG. 3-9). Repair vouchers are similar to service orders. They detail the work to be done, the name, address, and telephone number of the person requesting the work, a description of the item being worked on, and other necessary information.

Repair Voucher

Date______________________________

Time______________________________

Received of____

Address____

Phone number____

Item to be repaired____

Serial number____

Make____

Model____

Nature of problem____

Item accepted by____

3-9 Repair voucher.

Always include the serial number of the item being serviced on the voucher. Ideally, your repair voucher should have a questionnaire on it for the customer to fill out. The questions should inquire about the present condition of the item being left for repair. If you see a visible defect on the item, note it on the voucher, and have the customer sign to acknowledge the known defect. Repair vouchers offer good protection from disputes and accusations.

4

Organize your path to success

Organization is the cornerstone of all successful businesses. Without good organization, the best business ideas will be fruitless. Suppose you went to your accountant and found that your tax information had been lost. Would you continue to use the accounting firm? Doesn't it perturb you to go through a fast-food drive-through only to leave with an incorrect order? You see, when people are paying for your services, they expect you to perform in a professional manner. Good organization is a strong part of that professionalism.

TURN YOUR GOALS INTO ACHIEVEMENTS

We talked about setting goals in a previous chapter, here we are going to see how to achieve those goals. Setting goals is important, but achieving them is the key to success.

If you are not accustomed to working under a goal structure, start with some easy ones. For example, start with time management. Time management is a critical element to a successful business, so it is a good place to practice.

Build a work schedule. Set times for starting work, returning phone calls, paying bills, and so forth. Begin this exercise before you quit your current job. Follow your schedule for a week and see what happens. Keep a journal of your daily activities. Monitor your success. At the end of the week, see how your estimated time schedule coincided with your actual time needs. If you were on target with your schedule, congratulations!

If you find there are areas where you overestimated or underestimated your time needs, adjust your schedule and track another week of activities.

When you have mastered this simple schedule, make a more complex one and continue practicing. Your efforts will pay off when it counts.

Set aside time to read a certain amount of material on a daily basis. Whether it is 15 minutes or an hour, schedule reading time and use it to broaden your knowledge. There are a multitude of good books available to help new business owners in every facet of the business world. By setting and meeting reading goals, you practice self-discipline and learn how to build a better business.

Setting and meeting goals can be frustrating and stressful. The best way to avoid goal frustration is to set achievable goals. If you set your goals too high or move your goal sticks too quickly, you are going to become frustrated. If you want to run your own show, you must find effective ways to deal with frustration and stress. Setting and realizing realistic goals can help.

Goal achievement is not difficult. It is simply a matter of identifying a need or desire and accomplishing it. If you want to obtain a goal, you almost always can if the goal is realistic. Obviously, if your goal is to make a million dollars next week, the odds are against you. However, if you want to make a million dollars in the next 10 years, you might very well be able to do it.

SOLID ORGANIZATION EQUALS SMOOTH BUSINESS

The need for proper organization ranges from being able to find a pen to take down a phone message to starting your jobs on schedule. If you don't take the time to stay organized, your business will suffer. Let's look at some specific examples of how organizational skills and techniques affect your business and your profits.

Telephone organization

Most first impressions are made over the telephone. What does answering the phone have to do with organization? Organization plays a part in all business matters. When your phone is answered, what is said? If you simply answer with a quick "hello," callers will get the impression of a small business, probably one run from home. This isn't necessarily a bad impression, but it might not be the image you wish to present. On the other hand, if the phone is answered, "Good morning, Pioneer Plumbing, how may I help you," people will feel they are talking to a business with good business manners and a professional approach.

Your phone should be answered in a consistent manner. Switching back and forth between answering messages will imply that the business is unorganized and unprofessional. Pick a greeting and stick with it.

Once the phone is answered, how is the customer's request treated? If I call your business and ask to leave a message for you, what will happen? Am I going to be put on hold? Will the person I'm talking to have to scurry around to find a pen and paper to take down the message? If either of these events occur, I'm not going to be favorably impressed. Before you ever have the opportunity to return my call and talk with me, I've already formed an opinion and will be less receptive to what you have to say.

Will my phone call be returned promptly? When I leave my message, will I be advised when I can expect to hear from you? These issues are also a part of good organizational skills. Maybe you don't want to interrupt your field work to return calls during the day. That's fine, but tell me that I won't hear from you until after you are off the job. With the easy and affordable availability of pagers and cellular phones, there is little reason why you shouldn't be able to return my call during business hours. Perhaps you set aside specific times to return calls. Then callers will know when to expect you to return their calls.

What system do you employ to ensure you get all your phone messages? If you use an answering machine, you can be reasonably certain you got all your calls. But what if your secretary or spouse takes messages for you; can you be sure you got all the messages? Writing messages and phone numbers on scraps of paper is a poor way to run your telephone organization. Buy a phone log with duplicate pages and use it (FIG. 4-1). Write messages for service calls on a form that contains all necessary information and that can be given directly to your field personnel (FIG 4-2). Not only will using this type of message book reduce the chances of missed messages, it might come in handy for other reasons. For example, if Ms. Smith called two weeks ago and you need to follow up on the estimate you gave her, you can use the message book to retrieve her phone number. If you are forced into court over a customer dispute, the message book can serve as evidence. The usefulness of a message book far outweighs its cost.

If you have a professional answering service taking your calls, how will you track your messages? Most services maintain copies of all messages left and delivered, but you have no way of knowing if you got all of your messages. More importantly, it is difficult to determine how the service is treating your customers.

You should check up on the service from time to time. To do this you will need some help. The people working for the answering service will recognize your voice, therefore it helps to have friends leave messages for you. Monitor customer satisfaction and service courtesy by having your friends leave messages and by asking your customers how they are treated on the phone. By asking your customers, you not only get valuable information on the performance of your answering service, you make a great impression on the customers for being concerned about how they were handled on the phone.

Scheduling

Scheduling cannot be effective if you are not organized. It won't do any good to devise a perfect schedule if you can't find the schedule. Whether you are scheduling appointments for estimates or starting dates for jobs, you must be organized enough not to establish conflicting times. Once you have a good working schedule, you must be able to access it quickly and maintain it.

Many contractors use a scheduling board to keep track of their jobs. However, many of these same contractors fail to be as precise when they

Phone Log			
Date/Time	*Company Name*	*Contact Person*	*Remarks*

4-1 Phone log.

Service Call

Date____________________ Time____________________

Customer name____

Address____

Type of service requested____

Call taken by____

Call assigned to____

Service promised by____

4-2 Service call message form.

schedule estimates and meetings. If your business is doing well and you're busy, scheduling can become difficult. If you agree to meet a customer at a specific time, and don't make the meeting, you might lose a sale. At the least, you have made a bad impression on the customer.

My schedule is full of appointments, deadlines, and events. I use a bulletin board, a wipe-off wall calendar, a desktop daybook, and a briefcase appointment book to maintain my schedule. This may sound like overkill, but it works for me. I know my schedule at a glance. If I'm in the field, I refer to my briefcase appointment book to make on-the-spot decisions about my availability and requirements. If I'm at the phone in my office, I flip through the desktop daybook to schedule my work. When I walk into the office, I look up at the wall calendar and bulletin board for a quick assessment of my time needs. Using this system, very little slips between the cracks.

I'm not suggesting that you go to the extremes that I do. However, you must be able to know your time needs and responsibilities. Find a system that works for you, and use it.

Office organization

Office organization may be one of the most common organizational failures I find among my business clients. Usually, I can tell when I walk into a client's office if the business runs smoothly. If I look at a desk and see papers neatly stacked and placed in descending order along the desk, I'm impressed. This shows me that the client has his work organized and rated into priorities. On the other hand, if I see a desk covered in scattered papers, folders, and coffee cups, I jump to the conclusion that this client needs help getting organized.

Even if the cluttered desk is more productive than the neat desk, the visual impact on customers favors the well-organized work space.

Filing cabinets and in-out baskets are other ways to stay on top of your office organization. It is embarrassing to have a customer sit at your desk while you rummage through boxes, stacks of papers, or your briefcase looking for the estimate you presented last week. If customers see a lack of organizational skills in your office, are they going to have confidence in your ability to perform satisfactorily on the job? Would you want someone who couldn't find a recent estimate to plan the removal of your roof for a dormer addition? My point is this: be organized and professional at all times. Every contact you or your representatives have with customers will affect your business.

Field principles

To stay organized outside of your office, you need to establish and follow certain field principles. Have you ever come back to the office and been unable to find the deposit slip from the bank deposit you just made? When was the last time you went on an estimate and couldn't locate your notes about the job? If you are frequently frustrated because what you do in the field never seems to find its way back to the office, you need to set up some field criteria.

Most contractors practically live out of their trucks. It's easy to stick a phone number over the sun visor and forget where you put it. It is also entirely possible that the paper under the visor will be blown out of the truck and lost forever. To cut down on field losses, you need a mobile filing system.

A mobile filing system might consist of a portable filing box, the type sold in most department and office supply stores. A briefcase can serve as your rolling file cabinet. Once you determine what type of container will best hold your field records, set it up in your truck and use it.

In addition to needing an in-truck filing system, you should stock the vehicle with various office supplies. Paper clips, rubber bands, business forms, file folders, and related office supplies should be close at hand. I have found that a two-briefcase system works well for me. I stock one large briefcase with all the office supplies and small office equipment I may need. Then I use a smaller, less intimidating briefcase for my in-home appointments.

I am always prepared with the two-briefcase system. If my small briefcase is running low on estimate forms, I can restock it with forms from the larger in-truck briefcase. Once I've met with my customers, I set up a file for them in my truck. When I return to the office I make copies of the truck file and file them in the office. This way, if I'm caught on the road with the need to contact a recent customer, I have a copy of the customer's file with me. This dual filing system also reduces the odds of losing a file. If one of my files is lost, I have a duplicate in the truck. I admit this takes up space in the vehicle, but I'm always ready to do business.

GOOD BUSINESS FOLLOWS GOOD HABITS

Customers enjoy doing business with people who are organized and professional. You can eliminate much of your competition by acting like a professional. Let's look at two examples of how your organizational skills can help you get and keep more customers.

In our first example, the contractor is going to be unorganized. He is going to talk with Mr. Williams about a remodeling job. Mr. Williams wants to install a bathroom in his basement and finish off part of the basement for a family room. Our unorganized contractor, Ray, knows this is a job that will offer good work and great profit. Ray knows in advance what the estimate is and has plenty of time to prepare for it. Now, let's see how Ray handles the estimate.

On the day of the big estimate, Ray has been waterproofing the foundation of an addition he is building for a customer. His appointment with Mr. Williams is scheduled for 6:00 P.M. Ray leaves the waterproofing job at 5:00 P.M. He goes to a fast-food place for supper and washes up in the men's room. Ray eats his meal and heads over to see Mr. Williams.

On his way to the Williams' residence, Ray realizes he left the address of the property in his office. Ray thinks he remembers the house number, so he continues on his way. By the time Ray figures out that he is lost, it is getting close to 6:00 P.M. Ray gives up and searches for a public phone to call his office. He finds a phone and calls the office. His wife answers the phone and after digging around, finds Mr. Williams' address. Ray writes down the address and rushes to the appointment.

Ray arrives about 30 minutes late. Mr. Williams meets Ray at the front door. There is Ray, standing at the front door of a well-appointed home, dressed in torn jeans that are covered in waterproofing tar. Ray's shirt is dirty and his boots look like they have been through a war. Upon seeing Ray's condition, Mr. Williams asks Ray to meet him behind the house at the basement entrance. Ray comes into the basement and looks over the job. As Mr. Williams talks, Ray nods his head and looks around. When Mr. Williams finishes describing the work he wants done, Ray tells him that he will get back to him in a few days.

Ray leaves and Mr. Williams shakes his head. After going upstairs, Mr. Williams tells his wife about his meeting with Ray. The Williamses decide not to even consider Ray's bid. After looking at what they presently know about Ray, they wouldn't consider having him work in their home. What did Ray do wrong? Let's analyze some of his key mistakes.

Ray made many mistakes. First, he was late for the appointment. His tardiness could have been avoided with proper organization. Ray could have asked his wife to call the Williamses and explain that he was running a little late. Instead, Ray just shows up late, doesn't apologize, and starts off on the wrong foot.

Ray's personal appearance was a big mistake. Even though the tar had dried and the soles of his boots were clean, Ray looked as if he would be a major threat to the floor coverings, walls, and furniture in the house. His lack of proper dress eliminated any chance he had to sit down and talk with

Mr. Williams. Further, his appearance gave the impression of someone who has little respect for his customers or himself. This was not the image of a contractor that the Williamses wanted.

Ray made another mistake while listening to Mr. Williams. Ray did not take notes. Remember, this estimate was for extensive work. How could Ray remember all the details? Ray's lack of response to Mr. Williams and his choice not to take notes made a bad impression. Ray didn't offer any suggestions or advice to Mr. Williams. By saying nothing, Ray appeared uninterested in the job and possibly incompetent.

Ray left saying that he would get back to Mr. Williams in a few days. This left Mr. Williams not knowing when he would hear from Ray. If Mr. Williams had still been interested in dealing with Ray, he would have appreciated a specific date for the delivery of the estimate. Obviously, Ray blew this deal. Now, let's see how Craig handled the same estimate.

Craig is placed in the same circumstances that Ray had, but he does business in a very different manner. Prior to the day of the estimate, Craig rode by the home of Mr. Williams. This ride-by allowed Craig to locate the house and assess the neighborhood. Craig could see that the home was well-appointed and that Mr. Williams was a man who appreciated neatness.

On the day of the estimate. Craig left his waterproofing job at 4:00 P.M. He went home, showered, shaved, and dressed in clean clothes. He wore jeans, a flannel shirt, and boots, but they were clean and neat. Craig collected his photo album of similar jobs he had done, his notes, and briefcase. Craig left early for the appointment and was in the neighborhood with time to spare.

When Craig pulled into the driveway, Mr. Williams came to the door. Craig was right on time. The two men shook hands and Mr. Williams invited Craig into the living room. Craig spoke with Mrs. Williams and then they all went down to the basement.

As Mr. and Mrs. Williams described the work they wanted done, Craig took notes and made comments and suggestions. He made measurements and asked questions about the particulars of the job. After going over the plans for the basement, Craig and the Williamses went back up to the living room.

Sitting down, Craig showed the Williamses the photo album of his previous jobs. As conversation ensued, Craig got to know more about the Williamses. By the end of their talk, Craig was no longer a number from the phone book. He was a person with the possibility of becoming a friend.

As Craig was leaving, he told the Williamses that he would mail them letters of reference from his office the next morning. He explained that because of the size of the job, he would need a day to work up the estimate. Then Craig asked to meet with the Williamses on Saturday morning to go over the estimate with them, in case they might have questions about the job. The meeting was set and Craig left.

On Saturday, Craig met with the Williamses and went over the cost of doing the job. In less than two hours Craig had a signed contract and a deposit check. What did Craig do differently? It should be obvious: he was organized, neatly dressed, and professional in his business manners.

By wearing jeans and boots, Craig gave the appearance of someone who was not afraid to get involved in the physical work of the job, yet he didn't look like he spent all of his time in the trenches. By talking with the Williamses, Craig developed a rapport and gained the confidence of his customers. Talking allowed Craig to show his knowledge of the work required and his interest in the customer.

By taking measurements and making notes, Craig demonstrated his thoroughness in job planning and pricing. The photos Craig showed built a level of credibility. By offering to mail the Williamses reference letters, Craig took the first step toward overcoming an objection to closing the deal. By setting an appointment to go over the estimate, Craig created a setting for the signing of the contract. All in all, Craig did everything right, and much of what he did had to do with good organizational skills.

LEARN STRONG ORGANIZATIONAL SKILLS

Some people seem to have a natural ability to get and stay organized. Others seem never to be able to remember where they left their truck keys. Fortunately, good organizational skills can be learned. As a business owner, you owe it to yourself to learn as much as you can about effective organizational skills.

Where can you learn how to get organized? There are many ways to learn good organizational skills. You can read books, go to seminars, listen to cassette tapes, and meet with business consultants. Any of these methods can prove effective. The way that you choose to learn is not as important as the fact that you do choose to learn.

Many people are able to sit down and go through their daily activities in their minds. These people can trace their business requirements from start to finish. However, not all people have the ability to do this. For those people who have trouble visualizing their business structure and needs, a diary might be the answer.

If you keep a diary for a week, you can look through the pages and see what your week was like. You answered the phone, you returned phone calls, you went out on estimates, you dealt with customer complaints, and so on. By reviewing the contents of your diary, you can see what areas of your business need your attention.

Once you have an outline of your daily duties, choose methods of study that will enhance your day-to-day organization. Maybe you need to take a seminar that teaches effective time management. You might benefit from reading a book on professional secretarial duties. Whatever your weakness, work to improve it. The time and money you invest in getting organized will be returned to you many times over as you build your business.

5

Time management

Do you possess good time management skills? Do you understand the principles behind time management? If you are weak in this area, you are going to have some trouble in your business. This chapter is going to help you strengthen your skills in time management.

BUDGET YOUR TIME

Most people don't budget their time. They react rather than act. This is a major mistake for a business owner. If you compare it to a football game or boxing match, it is easy to see. The team that gets off the ball first or the man who throws the first punch has the advantage. The other side can only react. The same is true in business. Let me show you what I mean.

Consider this example. You are a remodeling contractor. You do much of your own work. Your strongest competitor is also a remodeling contractor, but she has an outside sales staff and uses subcontractors for her jobs. This leaves your competitor with more time to spend on business projections and evaluations. Both of you make about the same amount of money.

You try to be on the job by 7:30 A.M., and you rarely leave before 4:00 P.M. When you get out of the field, you take care of estimates and paperwork. By 6:30 P.M. you're home and having supper with your family. Then, from 8:00 to 10:00 P.M., you return phone calls, schedule subcontractors, and pay bills, among other things. By midnight you're in bed. This is a tough schedule to keep, but is not an uncommon one for small contracting companies.

While you thrash around and work yourself into exhaustion, your competitor seems to glide through life. She is in her office by 9:00 A.M., a full hour and a half after you are on the job. She handles her office work during the day, while you sweat it out in the field. She tends to customer service and supervision, while subcontractors do the heavy work.

Her commissioned salespeople make the estimate calls and sales. At 5:00 P.M. your competitor goes home. An answering service picks up her business calls and transfers them to the appropriate salesperson or worker. From 5:00 P.M. on, your competitor has a personal life. What are you doing wrong?

If you're happy, you're not doing anything wrong. If you resent your competitor, you need to work smarter, not harder. Both of you are operating profitable businesses, but you are working much harder. Your competitor has outmaneuvered you in the business arena. There are pros and cons to the life each of you live, but she seems to have the better life.

Doing business your way, you probably have less on-the-job problems because you are on the job. You are not paying out money to commissioned salespeople. You are in total control of your business. You are basically married to your business. You have chosen to create a working life, your competitor has created a business. Your competitor has her business running smoothly because she has made good management decisions. Part of her management is time management.

I started out having a working life, now I have a business. This is not to say I don't work. I probably work more hours than most business owners, but I enjoy most of what I do. For me, work is not a drudgery. I use subcontractors, sell my own work, do some of my own field work, do my own photography, writing, consulting, and real estate sales. As you can tell, I have diversified. I have variety in my work life. This variety allows for a change of pace and a more even keel. To accomplish this goal I've had to perfect my time-management skills.

RECOGNIZE WASTED TIME

It is easy to get caught up in the heat of the battle and lose your objectivity. You will not realize you are working harder and losing ground. To run a good business, you must be able to step back and look at the business operation from an objective point of view.

To determine how much time you are wasting, you may have to waste a little time. I realize this may seem redundant, but it's true. One of the best ways to pinpoint your wasted time is to spend time making a time log (FIG. 5-1). This log can expose how and where you are losing time.

Your log can be written or recorded on a tape recorder. As soon as you wake up, start your log. Keep track of everything you do from the time you wake up until you retire for the evening. This should include your business activity and your personal functions.

I know it might be inconvenient, but you must discipline yourself to make entries in your log for every activity you undertake. Whether it is brushing your teeth, walking the dog, going to the mailbox, or making business calls, enter your actions in the log. If you don't take the log seriously, this experiment will not work.

```
Appointments and Events for:   Feb 16, 1993

 6  AM   :
 7    :
 8    :  Meet with Mike about insurance
 9    :  9:25 car inspection
10    :
11    :
12  PM   :
 1    :
 2    :  Decision needed on blueprints for Walker job
 3    :
 4    :  Write memo about blueprint decision
 5    :
 6    :  Tennis
EVE   :  No appointments
```

5-1 Daily time log.

Keep your log for at least two weeks. At the end of that time, review the log page by page. Scrutinize your entries for possible wasted time. For instance, if you find you talk for more than 5 minutes on your business calls, take a close look at what you are talking about. Certainly there are times when business calls deserve 30 minutes or more, but most calls can be accomplished in 5 minutes or less.

Look for little aspects of your daily life that could be changed. Do you sit at the breakfast table and read the paper? How long do you spend reading the paper? Does this reading help you in your business? If your reading doesn't pertain to your business, maybe you should consider curtailing the time you spend at the table. If you derive pleasure from reading the paper, it's not a bad thing to do. However, if you are caught up in a habit of reading the paper for half an hour, and wouldn't feel deprived without this time, you have just found time to exercise.

You will probably find many red flags as you go through your time log. Almost everyone has little habits that rob them of time. Most of these routine activities have little impact on their daily life, but they continue because they are habits. Before you can change your bad habits, you must identify them. A time log will help you expose your wasted time.

REDUCE LOST TIME IN THE OFFICE

It is common to think that if you are in the office you're working. A lot of people sit in the office thinking they are working, without accomplishing anything productive. The people that have these feelings often lose more than time. They lose money.

Control employee gab sessions

Employees are supposed to make money for you, but they can drain you of your profits. If you spend too much time talking with employees, you lose money in two ways: you are not free to do your job, and your employees are talking to you instead of working. Let me give you a case history that will drive this point home.

During one of my consulting assignments, I was working with a service company to see how the company could improve its efficiency. The business owner was smart. He knew that if I was brought in and introduced as a consultant, the employees would not act as they normally did. For this reason, I was introduced as just another employee.

I acted like an employee, dressed like an employee, and got to know the employees. The first week I was inside the company, I found a major loss of income. The income loss was the result of upper management talking with employees.

When the crews would come into the office for their daily assignments, it was common for them to hang around for half an hour talking to the management. The talk was not business-related. There were 10 employees, an operations manager, and an office manager involved in the conversations. This company was charging $35 per hour for its labor rate. Based on billable time, the 10 employees wasting 30 minutes a day were costing the business owner $175 a day. The office manager and the operations manager were making a combined income in excess of $45,000 a year. When you tally up the cost to the business owner of these morning talks, the annual total of lost income was in the neighborhood of $45,500. What seemed like a simple, friendly morning talk was actually a business-threatening cash loss.

Schedule appointments for maximum efficiency

If you learn to set your appointments for maximum efficiency, you will see increased time in your day. You can convert this extra time into money. If you prefer, you can spend the time you save following your hobbies and being with your friends or family. In any event, setting efficient appointments will give you more time to use for the purpose of your choice.

Set efficient appointments by arranging meetings in a logical order. It is also beneficial to schedule meetings at your office, instead of in a customer's home or office. You will save time meeting in your office. However, if you are having a sales meeting you might be better off to sacrifice some time and meet with the potential customer on his home turf. People are more comfortable in their own home or office. When you are in a sales posture, you want the customer to be as comfortable as possible. But for now we are dealing with time management, so let's concentrate on why you want the meetings to convene in your office.

Set appointments in your office

How many times have you gone on appointments, only to have the other party be late or never show up? Have you ever thought of how much time these tardy or broken appointments cost you? Schedule appointments for your office, whenever possible, so if your client is late you can continue working. If the appointment is broken, you haven't lost any time.

There is a side benefit to meeting people in your office. You will be more at ease, and the people you are meeting will feel at a disadvantage. This can be detrimental in a sales meeting, but typically it will work to your advantage. Under these circumstances, you have control. How much you use or abuse it is up to you.

Monitor your activities

Do you sit in the office for hours at a time, waiting for the phone to ring? If you do, get a cellular phone or an answering service, and go out and look for business. If your phone isn't bringing you business, you must go out after it.

Do you take an hour to type a single proposal? If you do, it might be worth your while to find a typist that is an independent contractor. Hire the typist to transcribe your voice tapes into neatly-typed pages. This eliminates work you are not good at and allows you to do what you do best.

Assess your office skills and rate them. If filing is not your strong suit, find someone to do your filing. If you have an aversion to talking on the telephone, hire someone with an excellent phone presence to answer your calls. You can use a checklist to rate your in-office performance. After referring to your checklist, concentrate on the skills you need to improve.

REDUCE LOST TIME IN THE FIELD

Reducing lost time in the field is similar to reducing lost time in the office. You can use the same type of checklist to appraise your performance. However, the ways to improve the quality of the time you spend in the field will require some different tactics.

Use a tape recorder

You might be amazed at how a tape recorder can improve your efficiency. Many contractors spend an enormous amount of time driving. Tape recorders can turn this previously wasted driving time into productive time. Letters can be dictated, notes made, marketing ideas recorded, and a myriad of other opportunities preserved.

Tape recorders can improve efficiency in and out of the office. For an investment of less than $50, you can convert wasted time into productive time. This should result in extra money. Tape recorders are definitely worth strong consideration.

Should you have a mobile phone?

Cellular phones can be a boon to your business. Almost any owner of a service business can benefit from a mobile phone. If you have on-the-road communication, you can always be reached for emergency or highly-sensitive issues. When customers know they can reach you at any time, they are more comfortable doing business with you.

While it is true cellular phones are expensive on a cost-per-minute basis, that cost can be a bargain. If a $5 call results in a $10,000 job, you've made a wise decision. If a job has failed its footing inspection, you can call and cancel the concrete delivery. This can amount to a savings of at least $400.

A service technician can call ahead to make sure the next homeowner on the schedule is home and ready for service. If the homeowner has forgotten the appointment, the service technician can move up the schedule to the next service call. If you have a mobile phone and get stuck in traffic, you can call ahead and let the appointment you are going to know that you're going to be late. By keeping your customers and business associates aware of your schedule, you will have less broken appointments and disappointed clients.

Can you afford to operate without a mobile phone? With the progress in modern technology and the competitive nature of service businesses, I think cellular phones are nearly a necessity.

Eliminate lost time running to the supply house

Wasted time on the road is one of the most prevalent causes of lost income. Contractors that are not able to make accurate take-offs and schedule deliveries properly lose money on the road. When they or their employees leave the job for missing materials, profits are eroded. It may not be one of the larger financial losses a company will experience, but it is often the most regular loss of income.

During my time as a contractor and a consultant, I have witnessed countless situations when runs to the supply house pulled down the job profits. Most of these trips were avoidable. Even circumstances that are not reasonably avoidable can be made better with the use of logic. Let me give you a few examples of how running for materials will ruin your profits.

The first case history involves a plumbing and heating company. The job required the installation of hot-water baseboard heat. To install this type of heat, the mechanics needed copper tubing and copper fittings. All the work involved the same size pipe and fittings. The job was about an hour's drive from the plumbing shop and about 45 minutes from the nearest supply house.

On this particular morning, three mechanics and one helper were sent to the job. They knew what they would be doing and were supposed to have checked to make sure they had enough supplies. Since the mechanics were all seasoned veterans, their supervisor didn't pull the material for them, he just let them go on their own.

The crew had been on the job for a few hours before they called in on the radio. It seems that after getting into the job, they had discovered they were out of elbow fittings. These fittings were essential to the work being done. Neither of the two trucks on the job had enough fittings to finish out the day.

Obviously, for experienced mechanics to make such a mistake was alarming, but it got worse. If the shortage of fittings had been noticed earlier, the helper could have been sent to the supply house while the three mechanics continued to work. This would have involved a loss of about an hour and a half of road time, plus the time lost at the supplier.

What made matters worse was the timing of the discovery. No one realized the shortage of fittings until the last few were used. This meant no one could work until more fittings arrived. The company had three mechanics, at a billing rate of $35 per hour, sitting on their hands while a helper, with a billing rate of $25 an hour, went for a ride.

The obvious loss to the business owner was having to pay the crew to do nearly nothing. Then it got worse. While the three mechanics were sitting around, the general contractor arrived on the job. Finding three highly paid mechanics doing nothing infuriated the general. To make a long story short, the sub lost time, money, and the general contractor's future business.

All of this loss was the result of negligence. If the helper had been sent for fittings before the crew was at an impasse, the general contractor would not have been angry. If the supervisor had hand-held the mechanics, the loss could have been avoided. If the trucks had been properly stocked, the ordeal would not have occurred. There were many ways this financial disaster could have been avoided, but it wasn't.

In the second example, a young plumber was having trouble finishing his jobs on time. The business owner liked the employee, but couldn't afford the lost money in time that the plumber needed to accomplish routine tasks. I was brought in as a consultant to evaluate the employee.

I accompanied the employee and his helper to the job. I had been introduced as the new kid on the block, just another plumber. The job was roughing-in the plumbing for a new house. The job was nearly an hour away from the nearest supply house.

Almost before we started working, the plumber discovered he had forgotten a test-tee fitting. This fitting must be the first fitting installed on a plumbing stack. Seeing that he was missing the first fitting he needed, the plumber told the helper to go to the supply house and get one.

I intervened and suggested we start at a different point and see what else we might need, so as not to waste multiple trips to the supply house. The plumber agreed and we went about our work.

By noon, the list of needed items had grown to include four fittings. The plumber was again ready to send the helper to the supply house. I asked the plumber if he had looked over the remaining materials to determine if anything else was needed. He assured me he had. However, I recommended that they take a lunch break while I double-checked the material.

In looking over the job and the materials, I found a need for even more fittings and supplies. By the end of lunch, the list had grown to include several items. The helper went to the supplier and returned with the materials. The work was finished for the day, and I knew why the plumber couldn't meet his deadlines. He was inexperienced and couldn't look ahead. If I hadn't been on the job, there is no telling how many trips the helper would have made to the supply house, each time slowing down the job.

The findings of my report showed the employer how to improve the plumber's efficiency. The employee needed training in management, take-offs, and planning skills.

These types of problems go on everyday. As a business owner you cannot afford this type of wasted time. Except in rare cases, there is no suitable excuse for it. If you spend time preparing for a job, you will not waste time running for materials.

6

Initial financial considerations

Surviving the early years might be the toughest part of being in business for yourself. During these early years you establish business credit, learn the problems and pitfalls of owning your own business, and develop your customer base. If you can get past the first three years, you have a good chance of making it over the remaining rough spots.

EVALUATE YOUR CASH RESERVES

Any business needs cash reserves to get over the humps on the road to prosperity. A large number of businesses fail each year due to limited cash reserves. Without backup money, what will you do when a scheduled draw payment is held up and your bills are due? If you begin paying your bills late, your credit will be damaged.

Some people say you shouldn't start your own business until you have at least one year's salary in savings. I must admit that this would be a comfortable way to get started, but for most people, saving up a year's salary isn't feasible. When I first opened my plumbing business, many years ago, I borrowed $500 for tools and advertising. I had less than $200 in my savings account. Looking back, I was probably stupid to try such a venture, but I tried it, and it worked. I'm not, however, suggesting that you follow in my footsteps.

I am a risk-taker. For me, trying a new venture is an adventure. But I have learned never to gamble more money than I can afford to lose, and I think this line of thinking is good. Evaluate your responsibilities and situation to determine what, for you, are worthwhile gambles.

As you evaluate how much reserve money is enough for you, I suggest you play the worst-case game. Imagine scenarios of how your business venture might go. You are looking for what could be the worst possible outcome of your decision to be self-employed. Once you have said "what-if" enough, you will start to develop some insight into what you stand to lose.

Let's look at a quick example. You want to start a business as a tile installer. You have $3,000 in your savings account. Tools will cost $750. You already have a truck and the insurance company is willing to allow you to make monthly payments on your liability insurance. The economy is average. You feel you can get work with minimal advertising. You set an ad budget of $400 per month. You will be doing residential work and will get paid upon completion of your jobs. After making a personal accounting of your budget, you see that you need an income of $2,000 a month to help your spouse meet the bills. Do you have enough money to try jumping into business?

In my opinion, you don't have nearly enough cash to go into a full-time business. At the best, you have enough money to last one month, without income. Even if you are lucky enough to get work the first week, how will you pay for materials? What will you do if the customer doesn't pay you? Your odds of survival are very low under these circumstances.

However, if you keep your present job, you could start the business by working nights and weekends. This type of moonlighting is easy to schedule when you work small residential jobs. Homeowners like it because they can be home when the work is being done. As you do more jobs and your savings and customer base grows, you can consider turning the business into a full-time affair.

I don't think there is any clear-cut answer to how much reserve capital is enough. Each individual will have different needs. If I were forced to give an opinion, I would suggest having enough money to last at least four months without income. I would also suggest that when you feel you have established your monthly money need, add 20 percent to it. There will always be unforeseen expenses. Even with this reserve, you must monitor your success on a frequent basis. If the business is not going well or your budget spending is running high, look for alternative sources of income.

DECIDE HOW MUCH START-UP CAPITAL YOU'LL NEED

Start-up capital will probably be one of your first and most difficult hurdles. Start-up capital is the money you use to get your business off the ground. You will use it for advertising, office supplies, equipment, and other business-related expenses. This start-up money will go faster than you think. How much money do you need? The amount of money required will vary. The amount will depend on the type of business you start and how experienced you are in business matters.

It is best to rely on money you have saved for start-up capital. Borrowing money to start your business will put you in the hole right from the start. The burden of repaying a loan used for start-up money will only make establishing your business more difficult. However, many successful businesses are started with borrowed money.

When I started my first business, I had to borrow money just to buy tools. That first business was started about 18 years ago, with $500 to fill my toolbox. Today $500 doesn't seem like much money, but back then it felt like a big risk. I took the risk and it paid off. While I can't recommend going into debt to start your business, I did it and it worked.

A loan for start-up capital can be one of the most difficult to obtain. Lenders know that many new businesses don't survive their first year of operation. If you don't have home equity or some other type of acceptable collateral, banks will be reluctant to help you get your business started.

If you wait until you have quit your job to apply for a start-up loan, your chances of having the loan approved drop considerably. Most savvy entrepreneurs will arrange a personal loan while they have their regular job and use the funds for their business start-up. This procedure results in more loan approvals.

Some contractors start their businesses on a part-time basis. They are able to build up cash reserves and obtain the loans they need while they are still employed by someone else. The work they do on the side produces money that repays their loans and accumulates into a healthy stockpile of ready cash.

The stress and hours of working full-time and running a part-time business can be tiresome, but the results are often worth the struggle.

OPERATING CAPITAL NEEDS

Once your business is open, you will probably need some operating capital. This money is used to keep the business running until you receive job payments. If you don't already have a supply of money set aside for this purpose, you may want to apply for an operating-capital loan.

Lenders are a little more willing to make these loans, if you have an established track record. However, don't get your hopes up until you can provide two year's tax returns for income verification. Since you will probably need operating capital to survive your first two years, make arrangements for your financing before you jump into business.

Operating-capital loans can be financed in several ways. Some business owners borrow against their home equity to generate operating capital. Lines of credit are often set up for the contractor to pull from as money is needed. With this type of financing, you only pay interest on the money you use. Short-term personal loans are another way of financing your operating capital. These loans are frequently set up with interest-only payments until the note matures, at which time the total owed becomes due.

PICK LENDING INSTITUTIONS CAREFULLY

The first step toward establishing credit is picking your lending institutions. Not all lenders are alike. Some lenders prefer to make home mortgage loans. Others make car loans and other secured loans. Some lenders will make a loan for any good purpose, if they feel the loan is safe. Finding a bank that is willing to make an unsecured signature loan can be troublesome.

Before you set out to find a lender, decide what type of loan you need. If you are not a well-established and financially sound business, be prepared to sign the loan personally. Most lenders will not make a business loan without a personal endorsement. Are you looking for a secured or unsecured loan? A secured loan is a loan where something of value—collateral—is pledged to the lender for security against nonpayment of the loan. An unsecured loan is a loan that is secured by a signature, but not by a specific piece of collateral. Most lenders prefer a secured loan.

If you plan to request a secured loan, decide what you have for collateral. The amount of money you wish to borrow will determine the type and amount of collateral that is acceptable to the lender. For example, if you want to borrow $50,000 for operating capital, putting up the title to a $10,000 truck will probably not be sufficient.

When you are looking to borrow large sums of money, real estate is the type of collateral most desired by lenders. If you have equity in your home and are willing to risk it, you should find getting a loan relatively easy. However, home-equity loans can get confusing. Many people don't understand how much they can borrow against their equity. Let me show you two examples of how you might rate the loan value of the equity in your home.

Let's say your home is worth $100,000 and you owe $80,000 on the mortgage. Equity is the difference between the home's appraised value and the amount still owed on it. In this case, the equity amount is $20,000. Does this mean you can go to a bank and borrow $20,000 against the equity in your home? Most lenders would consider such a loan a high risk. Some finance companies might lend $10,000 on this deal, but they would be few and far between. In most cases, you would not be able to borrow any money against your equity.

Lenders want borrowers to have a strong interest in repaying loans. If a lender loaned you $20,000 on your equity, your house would be 100-percent financed. If you defaulted on the loan, you wouldn't lose any money, only your credit rating. To avoid this type of problem, lenders require that you maintain an equity level in your house that is above and beyond the combination of your first mortgage and the home-equity loan. Most banks will want you to have between 20 and 30 percent of the home's value remaining in equity, even after getting a home-equity loan.

Let's say you have the same $100,000 house, but you only owe $50,000 on it. A conservative lender might allow you to borrow $20,000 in a home-equity loan. This brings your combined loan balance up to $70,000, but you still have $30,000 equity in the home. If you default on this loan, you lose not only your credit rating, but $30,000 to boot. A liberal lender may allow you to borrow $30,000, keeping an equity position of $20,000.

If you don't have real estate for collateral, you can use personal property. Personal property might include vehicles, equipment, accounts receivable, certificates of deposit, or whatever. Different lenders will have different policies, so shop around until you find a lender you like.

Banks are an obvious choice when you apply for a loan. Commercial banks often make all types of loans. But banks are not your only option. Savings-and-loans normally deal in real estate loans, but they are worth investigating. If you belong to a credit union, check their loan policies and rates. Finance companies are usually aggressive lenders, but their interest rates may be high. Private investors are always looking for viable projects to invest in or to loan money to. Mortgage brokers are yet another possibility for your loan. A quick look through the phone directory and the ads in the local newspaper will reveal many potential loan sources.

HOW TO ESTABLISH CREDIT WHEN YOU HAVE NONE

This section is going to teach you how to establish credit when you have none. It is better to start with no credit than to start with bad credit. If you have never used credit cards or accounts, you will find it difficult to get even the smallest of loans from some lenders. Another problem you will encounter is the fact that you are self-employed. Most lenders don't want to loan money to self-employed individuals until the individuals can produce tax returns for at least two years. With this in mind, you might be wise to set up your accounts and credit lines before you quit your job.

Since most people don't set up business accounts until they are in business, I will give you advice on how to get credit without having a regular job. However, if you have the opportunity to establish your business accounts before you quit your job, do it.

As a new business you need credit, materials, and customers. As ongoing and competitive businesses, suppliers need new customers. In effect, you need each other; this is your edge. If you handle yourself professionally, you have a good chance of opening a small charge account. Now let's look at how you should go about opening your new supplier accounts.

You can request credit applications by mail or you can go into the stores and pick them up from the credit department. Once you have the credit applications, you will need some detailed information to complete them. Most applications will want personal references, credit references, your name, address, phone number, social security number, bank balances and account numbers, business name, and much more. The application will ask how much credit you are applying for. Don't write in that you are applying for as much as possible. Pick a figure, a realistic figure that is a little higher than what you really want. Setting the higher figure will give you some negotiating room.

Most supplier credit applications are similar. Once you fill out the first application, make copies of it for future reference. Not only will the photocopy serve to refresh your memory if the supplier has questions on your application, the copy will act as a template when you fill out other credit applications.

When you have completed the applications, return them to the credit department. You might want to hand deliver the applications. This will give you a chance to make personal impressions on the credit officers. Making good impressions on the credit managers might help to get you over the hump.

If you get a notice rejecting your credit request, don't give up. Call the credit manager and arrange a personal interview. Meet with the credit manager and negotiate for an open account. When all else fails, ask for a smaller credit line. Almost any supplier will give you credit for $500. This may not sound like much, but it's a start, and a start is what you need.

Take whatever credit you can get. After you establish the account, use it. Having an open account is not enough. You must use the credit to gain a good credit rating. If the credit is not used, it does you no good. You should use the credit account frequently and pay your bills promptly. When the supplier offers a discount for early payments, pay your bills early. By paying early and taking advantage of the discounts, you will save money and improve your credit rating.

Suppliers' accounts are the easiest way to build a good credit rating. After these accounts are used for a few months, your credit rating will grow. By keeping active and current accounts with suppliers you build a good background for bank financing.

Let's explore some supplier accounts where you may be able to establish credit accounts.

Advertising accounts

Advertising credit accounts are convenient. These accounts allow you to charge your advertising and pay for it at the end of the month. This eliminates the need to cut a check with every ad you place. It also gives your advertising a chance to generate income to pay for itself. Be careful—if you abuse these accounts, you can get into debt quickly.

Advertising often pays for itself, but sometimes it doesn't. You cannot afford to charge thousands of dollars worth of advertising that doesn't bring in paying customers. If you do, your business will be crippled before it starts. You will be faced with large payments on dud advertising. Use your advertising credit accounts prudently.

Supply stores

Since you will probably have to supply and install materials before your customers pay for them, supplier accounts buy you some float time, generally up to 30 days. You will be able to get the materials you need without paying for them until the first of the next month. This gives you time to install the materials and collect from your customers before paying for the materials.

Again, be cautious. If your customers don't pay their bills, you won't be able to pay yours. You must aggressively collect your accounts receivables.

Office supplies

Many times you can get better prices on office supplies by purchasing them from mail-order distributors. Having a business credit account with these distributors will make your life easier. You won't have to place COD orders and be concerned about waiting around to give the delivery driver a check. And you won't have to put business expenses on your personal credit cards. Business credit accounts simplify your accounting procedures.

Fuel

Paying cash for fuel for your trucks and equipment can be a real pain, especially when you have employees. Keeping up with the cash and the cash receipts is time consuming. By establishing a credit account with your fuel provider, your employees can charge fuel for your company vehicles and equipment. At the end of the month, your account statement will be easy to transfer into your bookkeeping system.

Vehicles and equipment

As your business grows, you may need more vehicles and equipment. Most business owners can't afford to pay cash for these large expenses. Financing or leasing vehicles and equipment will require a decent credit rating.

Major lenders

Major lenders are a little more difficult. They don't seem to feel the need to encourage business from new companies. Unless you have a strong credit rating, tangible assets, and a solid business plan, many banks will not be interested in loaning substantial sums of money. But don't despair, there are ways to work with banks.

Banks, like suppliers, should be willing to make a small loan to you. I know $500 will not buy much in today's business environment, but it is a place to start if you need to build a credit rating.

Bankers like to have collateral for loans. What better collateral could you give a banker than cash? You are thinking that if you had cash for collateral, you wouldn't need a loan. But, that is not always true. When you are establishing credit, any good credit is an advantage. Let me tell you how to get a guaranteed loan.

Set up an appointment to talk with an officer at your bank. Tell the banker you want to make a cash deposit in the form of a certificate of deposit (CD), but that after you have the CD on deposit, you want to borrow against it. Many lenders will allow you to borrow up to 90 percent of the value of your CD. For example, if you put $1,000 in a CD, you should be able to borrow about $900 against it. You are essentially borrowing your own money and paying the bank interest for the privilege. This concept may sound ludicrous, but it will help build your credit rating.

Banks report the activity on their loans to credit bureaus. Even though you are borrowing your own money, the credit reporting agency will show the loan as an active, secured loan. As long as you make the payments on time, you will get a good credit rating. This technique is often used by people repairing damaged credit, but it will work for anyone.

Building a good credit rating can take time. The sooner you start the process, the quicker you will enjoy the benefits of a solid credit history. The road to building a good credit rating can be rocky and tiresome, but it's worthwhile.

HOW TO OVERCOME A POOR CREDIT RATING

If you are starting with a bad credit rating, this section will give you options on how to overcome a poor credit rating. There is no question that setting up credit accounts will be harder if you have a poor credit history, but you can do it. If you are battling a bad credit report, plan on spending some time cleaning up the existing report and building new credit. This journey will not be easy, pleasurable, or quick, but the results should make you happy.

Secured credit cards

Secured credit cards are one way for people with damaged credit to begin the rebuilding process. Secured credit cards are similar to the procedure described for CD loans. You deposit a set sum of money with a bank or credit card company. Then you are issued a credit card. As the card holder, you can use the credit card with a credit limit equal to the amount of the cash deposit or slightly more. You are basically borrowing your own money, but you are rebuilding your credit.

CD loans

We have already talked about CD loans, so I won't go into extensive detail again. If you need to rebuild your credit, CD loans are a good way to do it. By depositing and borrowing your own money, you are able to build a good credit rating without risk.

Erroneous reports

Erroneous reports on your credit history are not impossible to fix. If you are turned down for credit, you are entitled to a copy of the credit report information used by the lender to make the decision to deny your credit request. If you are denied credit, immediately request a copy of your credit report. Credit reporting bureaus are not perfect, they make mistakes. Let me tell you a quick story about how my credit report was maligned when I applied for my first house loan.

When I applied for my first house loan, my request for the loan was denied. The reason for the denial was a delinquent credit history. I knew my credit was impeccable, and I challenged the decision. The loan officer talked with me and soon realized something was wrong. My wife's name is Kimberley and at the time of this credit request I didn't have any children. The credit report showed my wife as having a different name and it showed me having several children. Obviously, the report was inaccurate.

Upon further investigation, it was discovered that the credit bureau had issued the wrong credit report to my bank. My first and last name was the same as the person with the poor credit history. However, my middle initial was different, my wife's name was different, and I didn't have any children.

It happened that I lived on the same road as this other fellow. It was certainly a strange coincidence, but if I had not questioned the credit report, I would not have been able to build my first home.

I know from firsthand experience that credit reports can be wrong. I have seen various situations where my clients and customers fell victim to incorrect credit reports. If you are turned down for credit, get a copy of your credit report and investigate any discrepancies you discover.

Explanation letters

If your poor credit rating is due to extenuating circumstances, letters of explanation may help solve your credit problems. If there was a good reason for your credit problems, a letter that details the circumstances might be all it takes to sway a lender in your direction. Let me give you a true example of how a letter of explanation made a difference to one of my customers.

I had a young couple that wanted me to build a house for them. During the loan application process it came out that the gentlemen had allowed his vehicle to be repossessed. On the surface this appeared to be a deal-stopping problem. I talked with the young man and learned the details behind the repossession. At my suggestion, the man wrote a letter to the loan processor. In less than a week, the matter was resolved and the couple was approved for their new home loan. How did this simple letter change their lives? Let me explain.

This man had a new truck with high monthly payments. When he decided to get married, he knew he couldn't afford the payments. My customer went to his banker and explained his situation. The loan officer told the man to return the truck to the bank and the payments would be forgiven. However, the banker never told the man that this act would show up as a repossession on his credit report. My customer returned the truck with the best of intentions and acting on the advice of a bank employee.

When the problem cropped up and the bank employee was contacted, he confirmed my customer's story. The mortgage lender for the house evaluated the circumstances and decided the man was not an irresponsible person. It was decided that he had acted on the advice of a banking professional. Under the circumstances, my customer's loan was approved. If you have strong reasons for your credit problems, let your loan officer know about them. Once-in-a-lifetime medical problems could force you into bankruptcy, but they may be forgiven in your loan request. If you provide a detailed accounting that describes your reasons for poor credit, you may find that your loan request will be approved.

SEVEN TECHNIQUES TO ENSURE YOUR CREDIT SUCCESS

I am about to give you seven techniques that will ensure your credit success. This is not to say that these seven methods are the only ways to establish credit, but they are proven winners. Let's take a closer look at how you can make your credit desires a reality.

Get a copy of your credit report

In most cases, a written request will get you a copy of your credit report. This is a wise step to take before you try to establish new credit. By reviewing your credit report, you can straighten out any incorrect entries before you apply for credit.

Prepare a credit package

If you prepare a credit package before you apply for credit, you increase the chances of having your credit request approved (FIGS. 6-1 to 6-3). What should go into your credit package? If you own an existing business, your package should include financial statements, tax returns, your business plan, and all the normally requested credit information. If you have a new company, provide a strong business plan and the normal credit information.

Checklist of loan application needs

- ❑ Home address for the last five years
- ❑ Divorce agreements
- ❑ Child support agreements
- ❑ Social security numbers
- ❑ Two years tax returns, if self-employed
- ❑ Paycheck stubs, if available
- ❑ Employee's tax statements (i.e., W-2, W-4)
- ❑ Gross income amount of household
- ❑ All bank account numbers, balances, names, and addresses
- ❑ All credit card numbers, balances, and monthly payments
- ❑ Employment history for last four years
- ❑ Information on all stocks or bonds owned
- ❑ Life insurance face-amount and cash value
- ❑ Details of all real estate owned
- ❑ Rental income and expenses of investment property owned
- ❑ List of credit references with account numbers
- ❑ Financial statement of net worth
- ❑ Checkbook for loan application fees

6-1 Checklist of loan application needs.

Financial Statement

Your Company Name
Your Company Address
Your Company Phone Number

Date of statement: ______________________

Statement prepared by: ___________________________________

Assets		
Cash on hand	$ 8,543.89	
Securities	$ 0.00	
Equipment		
1992 Ford F-250 pick-up truck	$14,523.00	
Pipe rack for truck	$ 250.00	
40' Extension ladders (2)	$ 375.00	
Hand tools	$ 800.00	
Real estate	$ 0.00	
Accounts receivable	$ 5,349.36	
Total assets		$29,841.25
Liabilities		
Equipment		
1992 Ford F-250 truck, note payoff	$11,687.92	
Accounts payable	$ 1,249.56	
Total liabilities		$12,937.48
Net worth		$16,903.77

6-2 Financial statement.

PERSONAL FINANCES	Mar 1993		
Summary			**Over(Under)**
	Actual	**Budgeted**	**Budget**
Total Income	3140.43	2874.52	265.91
Total Expenses	2772.82	2749.87	22.95
Balance	367.61	124.65	242.96

Detail			**Over(Under)**
	Actual	**Budgeted**	**Budget**
Income			
Salary	2874.52	2874.52	
Other	265.91	0.00	Garage Sale
Expenses			
Withholdings			
Federal Income Tax	115.22	115.23	
State Income Tax	0.00	0.00	
FICA	90.88	90.88	
Medical	30.21	30.21	
Dental	22.90	22.90	
Other	0.00	20.00	Investment Fee
Total Withholdings	**259.21**	**279.22**	**(20.01)**
Percent of Budget	**9.02%**	**9.71%**	**-0.70%**
Finance Payments			
Credit Cards	33.80	33.80	
Auto Loan	238.50	238.50	
Home Mortgage	566.81	566.81	
Personal Loan	125.89	125.89	
Total Finance	**965.00**	**965.00**	**0.00**
Percent of Budget	**33.57%**	**33.57%**	**0.00%**
Fixed Expenses			
Child Care	0.00	20.00	
Property Tax	21.90	21.90	
Home Insurance	15.21	15.21	
Auto Insurance	80.44	80.44	
Life Insurance	102.55	102.55	
Contributions	310.00	286.00	
Vacation Savings	100.00	100.00	
Total Fixed	**630.10**	**626.10**	**4.00**
Percent of Budget	**21.92%**	**21.78%**	**0.14%**
Variable Expenses			
Household	54.80	50.00	
Groceries	395.66	425.00	
Auto Upkeep and Gas	37.10	45.00	
Furniture	0.00	15.00	
Clothing	35.90	30.00	
School	22.44	20.00	
Medical/Drugs	31.30	30.00	
Entertainment	36.88	45.00	
Memberships	10.00	10.00	
Cable TV	19.55	19.55	
Dining Out	38.40	30.00	
Gifts	100.00	50.00	
Pet Care	0.00	10.00	
Other	136.48	100.00	
Total Variable	**918.51**	**879.55**	**38.96**
Percent of Budget	**31.95%**	**30.60%**	**1.36%**

6-3 Sample budget.

Pick the right lender

Do some homework and find lenders that make the type of loans you want. Once you have your target lenders, take aim and close the deal.

Don't be afraid to start small

Don't be afraid to start small in your quest for credit. Any open account you can get will help you. Even if you are only given a line of $250, it's better than no credit line at all.

Use it or lose it

When you get a credit account, use it or you will lose it. Open accounts that are not used will be closed. In addition, an open account that doesn't report activity will not help you build your credit rating.

Pay your bills on time

Always pay your bills on time. Having no credit is better than having bad credit. If you have accounts that fall into the past-due category, your credit history will suffer and you will be plagued by phone calls from people trying to collect on their overdue statements.

Never stop

Never stop building your credit. The more successful you become, the easier it will be to increase your credit lines. However, don't get in over your head. If you abuse your credit privileges, it will not be long before you are in deep financial trouble.

7 Money management

Are you good at managing your money? Do you think you will be good at making the money in your business stretch to its maximum potential? If you don't learn to manage your money, the amount of money you make will have little bearing on your success. It is not a matter of how much money you make, but how much money you retain.

AVOID LARGE OVERHEAD EXPENSES

Overhead expenses can be enough to drive you out of business. Overhead expenses are expenses that are not related directly to a particular job. Examples of these expenses, called accounts payable, are: rent, utilities, phone bills, advertising, insurance, office help, and so on. If you fail to investigate and rate your overhead costs, you might find yourself looking for a job. Because this is such an important aspect of your business management, make a list of your accounts payable so you have them all listed on one sheet (FIG. 7-1). Now, let's take a closer look at how you can get a handle on your overhead expenses.

Rent

If you work out of your home, you will not notice a new financial strain when you convert one of your rooms into a designated office, but there are sacrifices. A home office doesn't get you away from your family and household. This can be a disadvantage, especially for people who lack self discipline.

If a home office won't cut it, you will have to turn to commercial space. With rent being an overhead expense, you will have to determine how much money you have to spend on rent and still maintain a desirable profit level.

Accounts Payable

Vendor	*Job*	*Amount Due*	*Date Due*	*Date Paid*

7-1 Accounts payable form.

Next decide how much space you need and what locations are acceptable to you. With this knowledge, shopping for an office is a simple matter. Respond to various for-rent ads and check prices and amenities.

Once you know what office space is renting for, you are in a position to make a decision. Some of the deciding factors might include the term of the lease, the amount of the security deposit, if you will pay for utilities or if they are included in the rent, and other related expenses.

For most contracting firms, a plush office is not necessary. As long as the office space is clean, well organized, and accessible, it should be fine. Keep your costs as low as you can, while still getting what you need, not necessarily what you want.

Utilities

Utilities are expenses you can hardly do without. These expenses include: heat, hot water, electricity, air-conditioning, public water fees, and sewer fees. You may be able to make minor cuts in these expenses, but you will have difficulty slashing the costs of these necessities.

Phone bills

Phone bills for a busy business can amount to hundreds of dollars each month. One way to reduce your expenses is to take advantage of all the discount programs offered by the many phone services. Another possibility is to make your long-distance calls after normal business hours. Reducing idle chit-chat can have a favorable impact on your phone bill.

Advertising

Advertising is a must-do expense for most businesses. However, you can make your advertising dollar stretch further by making your advertising more effective. Keep records on the pulling power of your ads by tracking the responses to your ads and the number of these responses that turn into paying work. Target your advertising to bring in the type of work that is the most profitable for your firm. Refine your advertising and it will pay for itself.

Insurance

You cannot reasonably avoid the burden of insurance. Most business owners resent this expense, but they can't afford to be without it. There are two keys to controlling your insurance expenses. The first key is to avoid over insuring your company. There is no reason to pay premiums for more insurance than you need. The second key is to shop rates and services. Insurance is a volatile market; the rates change often and quickly. The insurance you had last month may need to be reassessed this month. By doing periodic evaluations and shopping, you can maintain maximum control over your insurance expenses.

Professional fees

Professional fees might not be incurred on a monthly basis, but they can amount to hundreds, possibly thousands of dollars a year. Certainly you should engage professional help when you need it, but you can lower the cost of these services by doing some of the work yourself.

If you have a CPA do your taxes, and you probably should, you can save money by doing some preparation work. If you have all your documents organized and properly labeled, you will save the CPA time and you will reduce the number of phone calls, visits, and time you spend with the professional. The time you save the accountant will result in a lower fee.

Attorneys should be consulted on legal matters and they should draft and review legal documents. However, if you know what you want, draft an outline for the attorney. When you meet with your attorney, have your questions prepared, preferably in writing, and organized. The quicker you get in and out, the less you will have to pay.

Office help

Your business may or may not require office help. If you have personnel in your office, they are generally an overhead expense. Employees can be one of your most expensive overhead items. Don't generate this type of overhead expense until your business can't function properly without it.

Office supplies

Office supplies may not seem like a large expense, but they can add up. One way to reduce the cost of office supplies is to buy them in bulk. Instead of purchasing one legal pad, buy a case of pads. Buying in bulk and buying from wholesale distributors can reduce the money you spend on disposable office supplies.

Office furniture

Every office needs some office equipment and furniture. You will need a desk, a chair, and a telephone. You should have a filing cabinet, and you will want other pieces of furniture and equipment. Be selective in what you purchase. Before you buy anything, make sure you need it and that the cost is justified. Office equipment and furniture is an area where many business owners go overboard.

Copier

Do you need a copier? Every business has a need from time to time to make copies of documents. Some businesses do enough volume to justify buying or leasing a copier, but most small businesses don't. When you consider you can go to the local print shop and make copies for about 10¢ apiece, it will take a lot of copies to pay for owning or leasing a copier. Sure, going out to make copies is inconvenient, but it can save you a considerable amount of money.

Fax machine

Do you need a fax machine? Unless you deal with a large number of commercial clients you probably don't. If you use a fax infrequently, you can go to the print shop and pay a few bucks per page to have your documents faxed. Most of these pay-as-you-go fax places will allow you to use their fax number to receive incoming documents. When you consider that a fax will cost between $400 and $1,000 to purchase, you can send a lot of documents from the print shop before your investment is returned on a purchase.

Don't get caught up in the gadget trap. Many business owners like to buy gadgets for their offices. How often will you use a globe of the world? Do you really need a binding machine to bind your reports and proposals? Can you live without an electric stapler? Before you spend precious money on items that will do little more than be in your way, consider what you are buying.

Vehicles

Company vehicles can be considered overhead expenses. While it's true you need transportation, you don't have to have the ultimate in automotive engineering. If you can do your job in a $7,500 mini pick-up truck, don't buy a full-size truck at twice the price. Cutting your overhead is a matter of common sense and logic. Buy what you need and don't buy what you don't.

ANALYZE YOUR EXPENSES

Cutting the right expenses is one way to hone the edge on your profits. Cutting the wrong expenses may be worse than not cutting any expenses. You may lose more money than you save and some of your actions might be difficult to reverse. For these reasons, you must use sound judgment before you make cuts in your business expenses. We have just looked at normal overhead expenses. What other expenses might you cut back on?

Field supervisors

Field supervisors can be a significant expense. If you have a field supervisor, ask yourself if you could do the field supervision and use that individual as an income producer. Some business owners are unwilling to delegate duties to anyone, and others push too much responsibility off on employees. Field supervision is mandatory for many contractors, but unless there are numerous jobs going simultaneously, you might be able to supervise the field work.

There are many advantages to being your own field supervisor. You will see, firsthand, how your jobs are progressing. Customers will see you on the job and be more comfortable that they are getting a good job and special attention. The cost of having an employee as a supervisor will be eliminated or reduced. Before you pay high wages to a field supervisor, consider doing the work yourself.

Leftover materials

Leftover materials are common in the contracting business. Since the quantity of these leftovers is usually minimal, many contractors put the material into inventory. If the items will be used within a month, putting them into inventory is not a bad idea. However, if you don't know when you will have an occasion to use the materials, return them to the supplier for credit. Most suppliers that you regularly deal with will be happy to pick up your leftovers and credit your account. By taking this route, you are not paying to have the materials removed from the job site. When you need the materials again, the supplier will deliver them to you. You save money both ways.

Travel expenses

You can cut your travel expenses by keeping mileage logs for all your vehicles. When tax time comes, deduct the cost of your mileage from your taxes. The amount you may deduct is set by the government, but it is enough to make keeping a mileage log worthwhile.

Cash payments and receipts

Most contractors have occasions when they purchase small items with cash. The items might be nails, photocopies, stamps, or any number of other business-related items. Keep the receipts for these items and they become tax deductions (FIGS. 7-2 and 7-3). While a receipt for less than a dollar might not seem worth the trouble of recording, if you collect enough of them you will appreciate the savings. It is also important to record cash payments. Use a form that will give your customer a receipt and you a business record of the job and amount of payment (FIG. 7-4).

Directory advertising

Directory advertising in the phone book is one of the first expenses many business owners contemplate cutting. But before you make cuts in your directory advertising, remember that you will have to live with the change for a full year. If you reduce the size of your ad or eliminate it, you can't reverse your actions until the next issue is printed.

The size of your ad in the directory should be influenced by your competition. If all the electricians in town have big ads, you should probably have one, too. However, many people believe that a box ad in the column listings is sufficient for most types of businesses.

I am sure you will get calls from your directory advertising, but the cost of the ad could be more than the calls are worth. Track your calls for a year and decide if you are getting your money's worth from the phone book. You may be wise to have a modest listing in the directory and spend the money you save on a more targeted form of advertising. The main thing is to not make radical changes in your directory advertising before you are sure they are justified. A year is a long time to live with a mistake that hurts your business.

Cash Receipts

Date	*Account Description*	*Amount Paid*	*Date Received*

7-2 Cash receipts log.

Petty-cash record

Month____________________

Year______________________

Vendor	*Amount*	*Item*	*Date*	*Job*

7-3 Petty-cash record form.

Cash Receipt

Date______________________

Time____________________________

Received of____

Address____

Account number____

Amount received____

Payment for____

Form of payment____

Signed____

7-4 Customer receipt form.

Answering services

Human answering services are another frequent target of business owners looking to cut expenses. While some businesses do all right with answering machines, most businesses do better when a live voice answers the phone.

Most answering services can be terminated and picked back up the following month. If you are unsure of the value of your answering service, terminate it for a month and compare the number of leads you get for new work. If you don't notice a drop in business, you made a wise decision. If you lose business, reinstate the answering service.

Health insurance

Health insurance is very expensive, and many business owners consider eliminating their coverage at one time or another. You are taking a big risk if you drop your health insurance. As expensive as the insurance is, if you have a major medical problem the insurance will be a bargain. The accumulation of big medical bills could drive you into bankruptcy. If you feel you have to alter your insurance payments, look for a policy with a higher deductible amount. These policies will have lower monthly premiums and will still provide protection against catastrophic illness or injury.

Dental insurance

For most people, dental insurance is not as important as health insurance, but it is still a comforting thing to have. If you have bad teeth and are likely to need crowns and root canals, dental insurance can pay for itself. There are, of course, other types of dental services that will make your insurance premiums seem small. Try to avoid cutting out any insurance coverage.

Disability insurance

Disability insurance provides a buffer between you and financial disaster if you become disabled. You may get by without this type of coverage, but the gamble may not be worth the savings.

Inventory assets

Inventory assets often come under fire when money is tight. While you shouldn't carry a huge inventory that you don't need, you should stock your trucks with adequate supplies. If you cut back too far on inventory, your crews will waste time running to the supply house, and your customers will become frustrated by your lack of preparation for the job.

Retirement funds

Retirement plans are frequently one of the first expenses cut by contractors in a money crunch. A short moratorium on retirement funding is okay, but don't neglect to reinstate your investment plans before it is too late.

Bid-sheet subscriptions

Bid-sheet subscriptions are sometimes put under the financial microscope. While these expenses are not monumental, they appear to be an easy cut to make. If all you do with your bid sheets is glance at them and trash them, by all means cancel your subscription. If, however, you bid work on the sheets and win some jobs, eliminating your subscription could be the same as turning away work.

Credit bureaus

The fees charged by credit bureaus can become a target for company cuts. Before you make the decision to do without credit reports on your potential customers, weigh the risks you are taking. Doing work for one customer that doesn't pay would more than offset the savings you made by eliminating the credit bureau fees.

Sales force

When times are tough, business owners look at the sales force. Even when the salespeople are paid only on commission, some business owners consider eliminating them. This makes no sense to me. If you have a sales force that only gets paid for sales made, why would you want to get rid of them? Unless you are going to replace the existing sales force with new, more dynamic salespeople, the move to eliminate salespeople is senseless.

KEEP YOUR CREDIT ACCOUNTS UNDER CONTROL

Credit accounts can be the downfall of any business. If a business gets behind in its credit accounts, the business is likely to spiral downward. First will come late notices. Threats to turn the account over for collection will follow. If the situation is not rectified, lawsuits and judgments will come next. In the end, the business will be penniless and stuck with a bad credit rating. If you think starting a new business is a challenge, you will be staggered by the difficulty of bringing a wounded business up out of a hole.

Keep your accounts current. If you are unable to pay your bills, talk with your creditors. Credit managers are not ogres; they do have a job to do. If you are honest and open with your suppliers, most of them will work with you. If you try to ignore the problem, it will only get worse. A healthy business needs a good credit history. If you have good credit, cherish it; if you don't, work to get a good rating.

CASH FLOW IS PARAMOUNT TO YOUR SUCCESS

Paper profits are nice, but they don't pay the bills. Have you ever heard about the person who is land rich and cash poor? I've had tremendous financial statements and nearly no cash. Having a business with a high net worth is not worth much if you don't have enough money to pay the bills. Cash flow is very important to a healthy business.

I've seen a large number of businesses forced to the brink of bankruptcy, even though they had significant assets. These businesses held valuable assets, but couldn't convert the assets into cash. A business without cash is like an army without ammunition. The cash may be on the way, but if the enemy attacks before it arrives, the business cannot defend itself. Regardless of your assets and business strength, if you don't have cash, you are in trouble. As the old saying goes, "you've got to pay to play."

One of the biggest traps you should avoid is getting involved with bad jobs. Jobs that result in slow pay or no pay can be your undoing. Don't get greedy; greed is a major contributor to business failure. Most business owners are anxious to reach their goals quickly. It is better to take a slow approach and reach your goals than it is to run full out, only to fail.

CONTRACT DEPOSITS STRENGTHEN CASH FLOW

Getting money from your customer before you start a job is not always easy, but it does ease the cash-flow burden. It is not unusual for contractors who work with homeowners to receive cash deposits. A typical scenario finds the contractor getting one-third of the contract amount when the contract is signed, one-third of the contract amount at the halfway point of the job, and the balance upon completion. When this type of procedure is followed, you work with money provided by your customer. Without deposits, you must work with your own money and credit.

The public is very aware of the risks involved with giving money to someone before the work is done. This trend toward eliminating deposits will put more stress on your business. If you don't get an initial deposit, you must use your own credit lines and cash to maintain your business. Make arrangements now to be able to work with your own money and credit. Then carefully select your customers. If you are footing the bill up front, you have to be sure the customer will pay you. If you don't get deposits, you are at risk.

If you do get an up-front deposit, it is important that you use the deposit money for the job the deposit was made on. All too often contractors use deposit money to make truck payments, rent payments, or even to buy materials for other jobs. This is a potentially explosive proposition. You should never take one person's deposit and use it for someone else's job. You also should never use deposit money for ordinary operating expenses. If you get into this habit, you will probably be out of business in less than a year.

ELIMINATE SUBCONTRACTOR DEPOSITS

Eliminating subcontractor deposits is another way to wisely manage your money. Just as homeowners are reluctant to give deposits to contractors, you should be selective when giving deposits to subcontractors. Once a sub has your deposit money, it can be awful hard to get it back if the subcontractor doesn't perform as expected.

Accounts Receivable

Date	*Account Description*	*Amount Due*	*Date Due*	*Date Received*
	Total Due			

7-5 Accounts receivable form.

When subcontractors are dependent upon your deposit to do jobs, they may be in financial trouble. If you give in to these deposit requests, your money is at risk. Make it a rule to never give anyone money they haven't earned.

AVOID EXTENDING CUSTOMER CREDIT

While it is common for service companies to allow customers 30 days to pay their bills, this policy crunches your cash flow. People seem to have little problem neglecting to pay for services they receive and put on credit accounts. If you make an attempt to collect as soon as the job is done, most people will pay you and you won't risk losing your hard-earned money.

You might find that you must allow credit purchases to get business. But remember, working and supplying materials that you don't get paid for is worse than not working. If you allow credit purchases, be selective. Ideally, you should run credit checks on all customers who wish to establish a credit account. After all, will banks loan you money without checking your credit history?

COLLECT PAST-DUE ACCOUNTS

As a business owner, you will no doubt have occasion to collect past-due accounts. This part of your job can be frustrating. When you call to collect an old debt, you are likely to hear some very creative excuses. When you are listening to excuses for why you can't get the money owed you, don't get softhearted. Remember, if customers don't pay you, you cannot pay your bills.

If you start to acquire a long list of accounts receivable, be advised that you are headed for deep trouble. Never let debts get too old. As soon as a customer is in default, take action. Keep a readily available list of your accounts receivable and work with it until all accounts are paid (FIG. 7-5).

STRETCH YOUR MONEY

If you learn to stretch your money, you can do more business. The more business you do, hopefully the more money you will make. How can you stretch your money? Well, there are many ways to make your dollars go further. Let's see how you can increase the power of your cash.

- Rent expensive tools until you know you need them.
- Don't give subcontractors advance deposits.
- Collect job deposits from homeowners whenever possible.
- Make the best use of your time.
- Forecast financial budgets and stick to them.
- Buy in bulk whenever feasible.
- Pay your supply bills early and take the discount.

- Put your operating capital in an interest-earning account.
- File extensions and pay your taxes late in the year.
- Don't overstock inventory items.
- Keep employees working, not talking.
- Consider leasing big-ticket items.
- Study your business to find other ways to maximize your money.

8

Office space and location

For some businesses, office location is vital to success. However, most contracting businesses can function from a low-profile location. This is not to say that office space is not needed or is not important. Whether you work from your home or a penthouse suite, your office has to be functional and efficient, if you want to make more money.

FROM HOME OR A COMMERCIAL LOCATION?

I have worked from home and from commercial offices. My experience has shown that the decision to rent commercial space is dictated by your self discipline and the type of business you are running. Let's explore the factors you should consider when you think about where to set up shop.

Self discipline is paramount to your success when you work from home. If you are not able to make yourself stick to your work, you will find yourself out of business. It is easy to get caught sitting around the breakfast table for too long or taking a stroll around your farm. Working from home is very enjoyable, but you do have to set rules and stick to them.

Storefront requirements

Some businesses have storefront requirements. If you have a plumbing or an electrical business, you may need to display the fixtures you sell. If you need a storefront window, you probably can't work from home. But many contracting businesses don't require a showroom. Some examples of businesses that are not required to have a showroom include: general contracting, pest control, lawn care, carpentry, and so on. A storefront might improve business for any of these examples, but it is not mandatory. Storefront space is expensive. If you don't need it, why pay for it?

Home office

A home office is a dream of many people. Putting your office in your home is a good way to save money, if it doesn't cost you more than you save. Home offices can have a detrimental effect on your business. Some people will assume that if you work from home you are not well established and might be a risky choice. Of course, working from home doesn't mean your business is having financial trouble, but some customers are not comfortable with a company that doesn't have commercial office space.

I work from home now and I have worked from home at different times for nearly 20 years. I love it. I am also very disciplined in my work ethic, and even though my office is in the home, it is set in professional style. When clients come to my home office, it is obvious that I am a professional. I will talk more about setting up a home office a little later, but take your home office seriously.

Commercial image

A commercial office can give you a commercial image. This image can do a world of good for your business. However, the cost of commercial offices can be a heavy weight to carry. Before you jump into an expensive office suite, consider all aspects of your decision. We will talk more about the pros and cons of commercial offices as we continue with the chapter.

ASSESS YOUR OFFICE NEEDS

Before you can decide on where to put your office, you need to assess your office needs. This part of your business planning is instrumental to the success of your business. Can you imagine opening your business in a fancy storefront and then, say six months later, having to move out of the expensive rental space? Not only would that situation be potentially embarrassing, it would be bad for business. Once people get to know the location of your business, they expect it to stay there or to move up. A downward move, like the scenario described above, would alarm present customers and scare off a percentage of future customers. This is only one reason why office selection and location is important.

How much space do you need?

One of the first decisions you need to make before choosing an office location is how much space you will need. If you are the only person in the business, you may not need a lot of office space. The type of business you will operate from the office will have an effect on the size requirements. For example, a plumbing business can be run with minimal office space. Instead, storage space is needed for pipe, fittings, fixtures, and such. A heating or electrical company has about the same office requirements as a plumbing company. A general contracting firm may not need as much storage space, especially if the work is subcontracted out to others.

As you consider your space requirements, take the time to sketch a diagram of your proposed office space. It helps if you make the drawing to scale. How many people will you meet with at any given time? How many desks will be in the office? I'm now a one-man business, not counting subcontractors, but I have two desks and a sorting table in my primary office. In addition, I have another room designated as my library and meeting room, another space set up as a darkroom, and a photography studio in my basement. My barn stores my tools, equipment, and supplies. So you see, even a small business can need large spaces.

As you design your office, consider all your needs. Desks and chairs are only the beginning. Will you have a separate computer workstation? Do you need a conference table? Where will your filing cabinets go? Where and how will you store your office supplies? How many electrical outlets will be needed? The more questions you ask and answer before you make an office commitment, the better your chances of making a good decision.

Do you need commercial visibility?

Almost any business can do better with commercial visibility, but the benefits of this visibility may not warrant the extra cost. If you are out in the field working every day, and you don't have an employee in the office, what good will it do you to have a storefront? The woman looking for a replacement plumbing part is not going to come back to your store hours later. She is going to find a store that is open and has the part, unless you are the only game in town. If your business allows you to remain in the office most of the time, a storefront might be beneficial. You might get some walk-in business that you wouldn't get working out of your home. In general, if you can't be there to mind the store, you don't need the store.

Do you need warehouse space?

If you have to keep large quantities of supplies on hand or deal in bulky items, warehouse storage may be essential. This could be true for plumbers, carpenters, builders, and many other types of contracting businesses.

A popular solution for office and storage space is a unit that combines both under the same roof. These office/warehouse spaces are efficient, professional, and normally not outrageously expensive. If you don't need fast access to the materials you put into storage, renting a space at a private storage facility might be the best financial solution. At one time I ran a medium-sized plumbing company from a small office and a private storage facility. This arrangement was not convenient, but it was cost-effective and it worked.

LOCATION MAKES A DIFFERENCE

If you provide in-home service to customers in the city, living in the country might be inconvenient and might cause you to lose customers, especially if you charge extra for travel time. Having an office where you are allowed to display a large sign is excellent advertising and builds name recognition for your company. If your office is in a remote section of the city, people may not want to come to your office. If you work from your home, and your home happens to be out in the boonies, customers may not be able to find you, even if they are willing to try. Location can make a difference in the profitability of your business.

OFFICE LOCATION AFFECTS YOUR PUBLIC IMAGE

The public can be a strange group. Public opinion is fickle, but important. If your business is perceived to be successful, it probably will become successful. On the other hand, if the public sees your business as a loser, look out. It is unfortunate that we sometimes have to make decisions and take actions just to create an image the public wants to see, but there are times when we must.

What does the location of your office have to do with the quality of your business? It probably has nothing to do with it, but the public thinks it does. For this reason, you must cater to the people you hope will become your customers. If you are dealing with a business where a downtown office is expected, you should plan on working your way into a downtown office. If you don't, a time will come where your customer base will peak and stagnate. There are only so many people that will deal with you when your business is unconventional.

Aside from the prestige angle of office location, you must consider the convenience of your customers. If your office is at the top of six flights of steps, with no elevators available, people may not want to do business with you. If there isn't adequate parking in the immediate vicinity of your office, you might lose potential customers. All of these location factors play a part in your public image and success.

HOW MUCH OFFICE CAN YOU AFFORD?

When you look at your budget for office expenses, you must consider all the costs related to the office. These costs might include: heat, electricity, cleaning, parking, snow removal, and other expenses. These incidental expenses can add up to more than the cost to rent the office.

If you rent an office in the summer, you might not think to ask about heating expenses. In Maine, the cost of heating an office can easily exceed the monthly rent. When you calculate your office budget, take all related expenses into account. Come up with a budget number that makes you comfortable, and make sure you keep your office rent and related expenses within your budget.

When you begin shopping for an office outside of your home ask questions, and lots of them. Who pays for trash removal? Who pays the water and sewer bill? Who pays the taxes on the building? These may seem like stupid questions, but some leases require you, the lessee, to pay the property taxes. Who pays the heating expenses? Who pays for electricity? Who pays for cleaning the office? Does the receptionist in the lower level of the building cost extra? If there is office equipment in a common area, like a copier for example, what does it cost to use the equipment? Your list of questions could go on and on.

Ask all the questions and get answers. If you are required to pay for routine expenses, like heat or electricity, ask to see the bills for the last year. These bills will give you an idea of what your additional office expenses will be.

Before you rent an office, consider the ups and downs of your business cycle. If you are in a business that drops off in the winter, will you still be able to afford the office. Do you have to sign a long-term lease, or will you rent on a month-to-month lease. It usually costs more to rent on a month-to-month lease, but for a new business a long-term lease can spell trouble. If you sign a long lease and default on it, your credit rating will be scarred. What happens if your business booms and you need to add office help? If you are in a tiny office with a long-term lease you've got a problem. If you do opt for a long-term lease, negotiate for a sublease clause that will allow you to rent the office to someone else if you have to move.

It can be easy to fantasize about how a new office will bring you more business. It's fine to enjoy this thought, but don't trap yourself. When you project your office budget, base your forecast on your present workload. Better yet, if you've been in business awhile base the projections on your worst quarter for the last year. If you can afford the office space in the bad times, you can afford it. If you can only afford it during the summer boom, you're probably better off without the office.

Many new business owners get carried away with their offices. They look for space with marble columns, fancy floors, wet bars, and all the glitter depicted in offices on television. Well, unless you are independently wealthy, these lavish work spaces can rob you of your profits. Not only will the cost of the office drain your cash flow, you may lose business because of your expensive taste. That's right, you may lose business by renting a great office.

Consumers aren't stupid. It doesn't take long to figure out that if a company has high office overhead, the customers are paying for it. The flashy office may be fun, and it may be impressive, but it can also be bad for business. Of course, this will depend on the type of business you have and your clientele, but don't assume that an expensive office is going to get you higher net earnings.

ANSWERING SERVICES vs. MACHINES

When the pros and cons of answering services are compared to answering machines, you may find many different opinions. Most people prefer to talk to a live person, rather than a cold, electronic machine. However, as our lives become more automated, the public is slowly accepting the use of electronic message storage and retrieval.

Which do you prefer, an answering service or an answering machine? Do most of your competitors use machines or live people to answer their phones? This is easy to research, just call your competitors and see how the phone is answered. It's pretty well accepted that the use of an answering machine might cause you to lose business, but that doesn't necessarily mean that you should not consider the use of a machine.

Answering machines are relatively inexpensive. Most machines are dependable. These two points give the answering machine an advantage over an answering service. Answering services are not cheap, and they are not always dependable. More callers will leave a message with an answering service than would on a machine. This point goes in favor of the answering service. Answering services can page you to give you important and time-sensitive messages; answering machines can't. Another point for answering services. Are you confused yet? Don't worry, we'll sort it all out.

To determine which type of phone answering method you should use, make a list of the advantages and disadvantages of each. Once you have your list, it will be easy to arrive at a first impression on which option you should choose. It may be necessary to change your decision later, but at least you will be off to a reasonable start.

Answering machines

There are several qualities you should look for in an answering machine. These qualities should include the ability to check your messages remotely. Most modern answering machines can be checked for messages from any phone. Choose a machine that allows the caller to leave a long message. Many machines will allow the caller to talk for as long as they want. These machines are voice activated and will cut off only when the caller stops speaking.

Pick a machine that will allow you to record and use a personal outgoing message. Some of the answering machines are set up with a standard message that you can't change. It is beneficial to customize your answering message. If you buy an answering machine that meets all of this criteria, you should be satisfied with its performance.

Answering services

Price is always a consideration, but don't be too cheap. You may get what you pay for. Find an answering service that answers the phone and takes messages in a professional manner. You will want a dependable service, one that will see that you get all of your messages. Ask if you can provide a script for the operators to use when answering your phone. Some services answer all the phones with the same greeting, but many will answer your line any way you like.

Ask what hours of the day you will be getting coverage. Most services provide 24-hour service, but it generally costs extra. Ask if the service will page you for time-sensitive calls—most will. Determine if your bill will be a flat rate or if it will fluctuate based on the number of calls you get. Inquire about the length of your commitment. Some answering services will allow you to go on a month-to-month basis; others want a long-term commitment.

If you decide to use an answering service, periodically check on their performance. Most services provide a special number for you to call to pick up your messages, especially if they base your bill on the number of calls taken on your phone line. Even if you have to pay for calling in on your own line, do it every now and then. When you call in on your business number, you can get firsthand proof of how they handle your calls. Have friends call and leave messages. The operators won't recognize the voices of these people and will treat them like any other customer. This is the best way to check the performance of answering services.

Which should you choose?

If your callers expect fast service, like service plumbers and electricians, go with an answering service. When people with emergency repair needs call and get a machine, they will call around to find another contractor. If you have a business that doesn't require immediate response, such as lawn care, a machine may serve your needs. You will lose some business, but you will also save some money.

I think the business you lose with an answering machine is more valuable than the money you save. If you can hire a human answering service, I think you should. I have tried having my phones answered each way, and I am convinced that human answering services are the best way to go.

9 Computers and software

Computers and software have made a monumental impact on the way modern business is conducted. With the use of computers and the associated peripherals that go with them, a business can do almost any office task. Modems can be used to send documents over phone lines from one computer to another. Credit checks can be flashed on a monitor in a matter of moments. Bookkeeping can be done by people who before would never have attempted such a task. Scanners can put hand-drawn sketches into the computer for modification. Estimating programs can take the guesswork out of job quotes. Computers are truly remarkable.

As good as computers are, they are only as good as the individuals who use them. If the person operating the machine doesn't know how to use it, the job will not get done or will be done incorrectly.

Many people are intimidated by computers. This group of people prefer to muddle along the old-fashioned way rather than learn new skills. They will present arguments about the length of time they will need to become computer-literate. Money will be another defense. Excuses will be given for not being able to afford a computer system, even though many systems can be purchased for less than $1,000. Some people will never learn to benefit from modern technology. When you compete against this group of people, you can gain a competitive edge by using computers.

COMPUTERS ARE A WAY OF LIFE

Most businesses that have used computers for any length of time wouldn't know how to operate without them. Once you are bitten by the computer bug, the attraction can be consuming. It starts with a word-processing software package. You see how much easier it is to prepare your contracts and correspondence on the computer (FIGS. 9-1 to 9-7). You no longer need correction tape for the old typewriter. Having forms saved on your computer disk makes it fast and easy to turn out proposals, letters, and much more.

The next step is spreadsheet software. You start playing the what-if game and get excited about the potential of your company. Suddenly, forecasting the future and tracking your budget is fun. You are starting to get hooked.

Then you experiment with database programs. Mailing lists were never so easy to accumulate and use. Your marketing will be much easier with your new, computerized customer base. Checking inventory is a snap, and storing historical data is a breeze.

With a little more time you get into accounting programs and automated payroll. With a little playful study you find you can do what you've been paying other people to do for you. For the one-time cost of your software, you have eliminated a routine overhead expense.

Sooner or later you give in to the games. Playing golf on the computer may not be the same as kicking the dew off grass at the club, but it's not a bad way to relieve stress. By this time the computer bug has you in its grip. You are addicted and would rather give up your easy chair than your computer.

If you don't like the idea of using computers, you may find this scenario hard to believe, but don't be surprised to find yourself in a similar situation.

I used to hate computers. I didn't believe in them and wanted nothing to do with them. My wife, on the other hand, was fascinated with computers. When she wanted to buy one, I didn't object, but I showed no interest in the mechanical monster. Kimberley went about her business with the computer. She often tried, unsuccessfully, to get me to use it. When Kimberley wanted to computerize our plumbing and remodeling business, I was adamant it would never happen. Well, I was wrong.

We did put out business on the computer. After awhile I started to see the benefits of the changeover. The office requirements were met much more quickly than before. At first, Kimberley handled all of the computerized tasks. In time I started to play with the system. I have never liked having a system, whether it be manual or computerized, that I couldn't understand. At first I was frequently frustrated. But after a short time I started getting into the new technology. Soon I insisted we upgrade our system to a more modern and powerful level. The rest is history. I now know and love computers.

WHY DO YOU NEED A COMPUTER?

You probably don't need a computer, but if you have a computer you have many advantages. A computer will save you time. Your work will be less tedious. Chances are good that your business will be better organized and more efficient with a computer. There are lots of reasons to buy a computer and very few reasons not to.

Computers give you an edge in marketing your services. Once your computer system is on-line, you can, and probably will, perform marketing ploys you have never done before.

Your Company Name
Your Company Address
Your Company Phone Number

Quote

This agreement, made this _____________ day of ____________ , 19______, shall set forth the whole agreement, in its entirety, by and between Your Company Name, herein called Contractor and___________, herein called Owners.

Job name: ____

Job location: ____

The Contractor and Owners agree to the following:

Contractor shall perform all work as described below and provide all material to complete the work described below. Contractor shall supply all labor and material to complete the work according to the attached plans and specifications. The work shall include the following:

____

____

____

Schedule

The work described above shall begin within three days of notice from Owner, with an estimated start date of _______________. The Contractor shall complete the above work in a professional and expedient manner within ___ days from the start date.

Payment Schedule

Payments shall be made as follows:

____

____

This agreement, entered into on _____________, shall constitute the whole between Contractor and Owner.

____________________	____________________
Contractor Date	Owner Date
	____________________
	Owner Date

9-1 Work quote.

Your Company Name
Your Company Address
Your Company Phone Number

PROPOSAL

Date: ______________________________

Customer name: ____

Address: ____

Phone number: ____

Job location: ____

Description of Work

Your Company Name will supply, and or coordinate, all labor and material for the above referenced job as follows:

____

____

____

Payment Schedule

Price: ____($__________),

Payments to be made as follows:

____

____

____

All payments shall be made in full, upon presentation of each completed invoice. If payment is not made according to the terms above, Your Company Name will have the following rights and remedies. Your Company Name may charge a monthly service charge of one-and-one-half percent (1.5%), eighteen percent (18%) per year, from the first day default is made. Your Company Name may lien the property where the work has been done. Your Company Name may use all legal methods in the collection of monies owed to it. Your Company Name may seek compensation, at the rate of $____________ per hour, for attempts made to collect unpaid monies.

Page 1 of 2 initials __________

9-2a Job proposal (page 1).

Your Company Name may seek payment for legal fees and other costs of collection, to the full extent the law allows.

If the job is not ready for the service or materials requested, as scheduled, and the delay is not due to Your Company Name's actions, Your Company Name may charge the customer for lost time. This charge will be at a rate of $_______ per hour, per man, including travel time.

If you have any questions or don't understand this proposal, seek professional advice. Upon acceptance, this proposal becomes a binding contract between both parties.

Respectfully submitted,

Your name and title
Owner

Acceptance

We the undersigned do hereby agree to, and accept, all the terms and conditions of this proposal. We fully understand the terms and conditions, and hereby consent to enter into this contract.

Your Company Name	Customer
By ______________________	______________________
Title______________________	Date______________________
Date______________________	

Proposal expires in 30 days, if not accepted by all parties.

9-2b Job proposal (page 2).

With the automated features offered by your computer, you can experiment and find new ways of getting business. Creating mailing lists will be simple. Designing your advertising will be fun. Tracking past successes and failures will allow you to refine your marketing plan. In general, computers are a great aid in building your business.

COMPUTERS CAN HELP YOUR BUSINESS

A computer can help your business in a myriad of ways. You can perform almost any clerical or financial function on a computer. You can draw blueprints on a computer. Marketing and advertising will be easier when you utilize a computer. Inventory control and customer billing will almost take care of themselves on a computer. Let's take a closer look at exactly how you can derive benefits from computerizing your office.

Certificate of Completion and Acceptance

Contractor: ____

Customer: ____

Job name: ____

Job location: ____

Job Description: ____

____

____

____

Date of completion: ____

Date of final inspection by customer: ______________________________

Date of code compliance inspection & approval: ______________________

Defects found in material or workmanship: __________________________

____

Acknowledgment

Customer acknowledges the completion of all contracted work and accepts all workmanship and materials as being satisfactory. Upon signing this certificate, the customer releases the contractor from any responsibility for additional work, except warranty work. Warranty work will be performed for a period of one year from the date of completion. Warranty work will include the repair of any material or workmanship defects occurring between now and the end of the warranty period. All existing workmanship and materials are acceptable to the customer and payment will be made, in full, according to the payment schedule in the contract, between the two parties.

______________________________ ______________________________

Customer Date Contractor Date

9-3 Certificate of completion and acceptance.

Your Company Name
Your Company Address

Dear Sir:

I am soliciting bids for the work listed below, and I would like to offer you the opportunity to participate in the bidding. If you are interested in giving quoted prices on material for this job, please let me hear from you, at the above address.

The job will be started in _______ weeks. Financing has been arranged and the job will be started on schedule. Your quote, if you choose to enter one, must be received no later than __________________________.

The proposed work is as follows:

____

____

____

Plans and specifications for the work are available upon request.

Thank you for your time and consideration in this request.

Sincerely,

Your name and title

9-4 Form letter for soliciting material quotes.

Marketing

Marketing can be done more efficiently with a computer. You can build a database file of your current and potential customers. Then you can print mailing labels for all these names. If you use telemarketing, you can use computers to conduct your cold-calling surveys. Drawing programs will enable you to create logos and flashy flyers, while cutting down on your design and typesetting costs.

Following historical data is easy with a computer. With just a few keystrokes you can see what decks were selling for two years ago. Tap the keys a few more times and the display will tell you the type of work from which you have made the most money. Computers can definitely make your marketing more effective and faster.

Your Company Name
Your Company Address

Dear Sir:

I am soliciting bids for the work listed below, and I would like to offer you the opportunity to participate in the bidding. If you are interested in giving quoted prices for the Labor/Material for this job, please let me hear from you at the above address. The job will be started ___________. Financing has been arranged and the job will be started on schedule. Your quote, if you choose to enter one, must be received no later than ___________________________.

The proposed work is as follows:

____

____

____

____

Thank you for your time and consideration in this request.

Sincerely,

Your name and title

9-5 Form letter for soliciting bids from subcontractors.

Payroll

You will enjoy the automated features available on payroll software. By entering a minimal amount of data, the computer will run your payroll for you. This is a real timesaver for companies with several employees.

Job-costing results

Job-costing results can be printed out of your computer files in minutes. What used to take hours to do can now be done in minutes. Not only will computer-generated job-costing numbers come more quickly, they will probably be more accurate.

Promissory Note

City ______________________________

State ______________________________

Date ______________________________

Face amount of note $________________

For value received and/or services rendered, the undersigned promises to pay to the order of ____________________ at ____________________ the principal sum of $____________ ($__________) with interest thereon at the rate of ____________ percent per annum, said interest to be paid in monthly payments of $__________, ($__________) for __________ months. The balance is due in full and payable on __________. This note shall be secured by the personal guarantee of the undersigned and their heirs. Further security for this note shall be as follows:

____

____

________________________   ________________________

Debtor Date   Debtor Date

____

Witness Date

9-6 Promissory note.

Track your budget

Tracking your budget is no problem with the right software. You can boot up the computer and see where you stand at any time during the year (FIGS. 9-8 and 9-9). This is not only convenient, it can keep you from busting your budget.

Project tax liabilities

Projecting tax liabilities is another function a computer can perform for you. If you are wondering how much money you will need at tax time, just ask the computer. Without hesitation, the computer will give you the information you need to plan for your tax deposits.

Balloon Promissory Note

City ______________________________

State ______________________________

Date ______________________________

Face amount of note $____

For value received and or services rendered, the undersigned promises to pay to the order of _______________ at _____________ the principal sum of $________________ ($_______________) with interest thereon at the rate of __________ percent per annum, said principal and interest to be paid in full on __________. This note shall be secured by the personal guarantee of the undersigned and their heirs. Further security for this note shall be as follows:

____

____

____

______________________ Debtor Date	______________________ Debtor Date
______________________ Witness Date	

9-7 Balloon promissory note.

Computerized estimating

Computerized estimating will make your life easier. There are systems available that will almost do the estimates for you. Of course you have to give the machine a little help, but not much. By moving a pen across the blueprints, you can have take-offs done before you could get started manually.

Word processing

Word processing will make all of your written work easier. With built-in spell checkers and thesauruses, today's word-processing software takes the drudgery out of writing a letter. If your grammar skills are rough around the edges, you can incorporate grammar software to correct your writing.

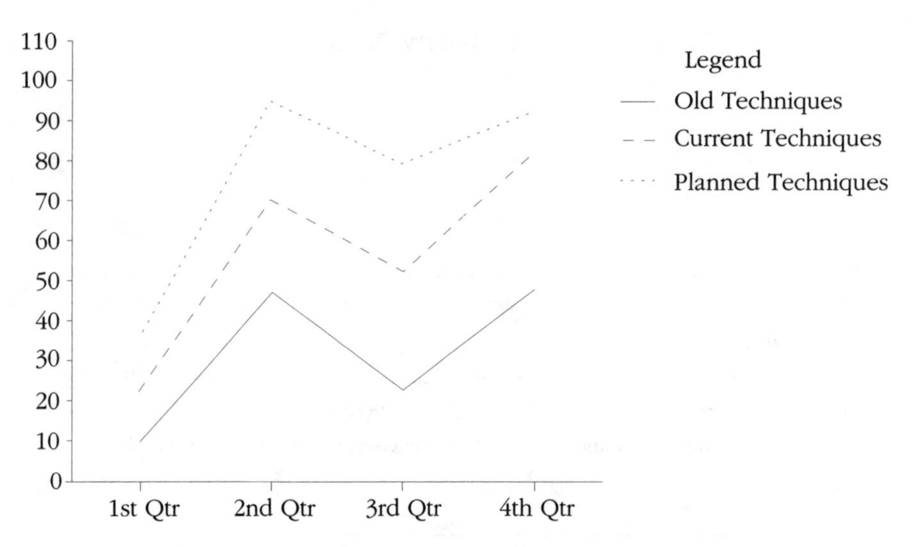

9-8 Sample tracking chart that can be created with a computer.

Databases

Databases can be used to store information on any subject. Whether you want to know the cost of a ton of topsoil or when your trucks are due for service, a database can do the job. You can design the database file to include as much or as little information as you like. Sorting these electronic files is much easier than digging through the old filing cabinet.

Checks and balances

If you don't like writing checks or keeping track of your bank balance, let the computer do it for you. There are dozens of programs and forms available to help you pay your bills with a computer (FIG. 9-10). As a side benefit, the computer will make adjustments to your bookkeeping records as it goes.

Customer service

Customer service can be improved with the use of a computer. If you receive a repair call, you can quickly determine if the job is still covered under warranty. When your customers are having a birthday, your computer can remind you to send out a birthday card. If your customers require routine maintenance, like recharging their air-conditioning, cleaning their furnace, or charging their water-treatment equipment, the computer can send out reminders to the customers of their service needs.

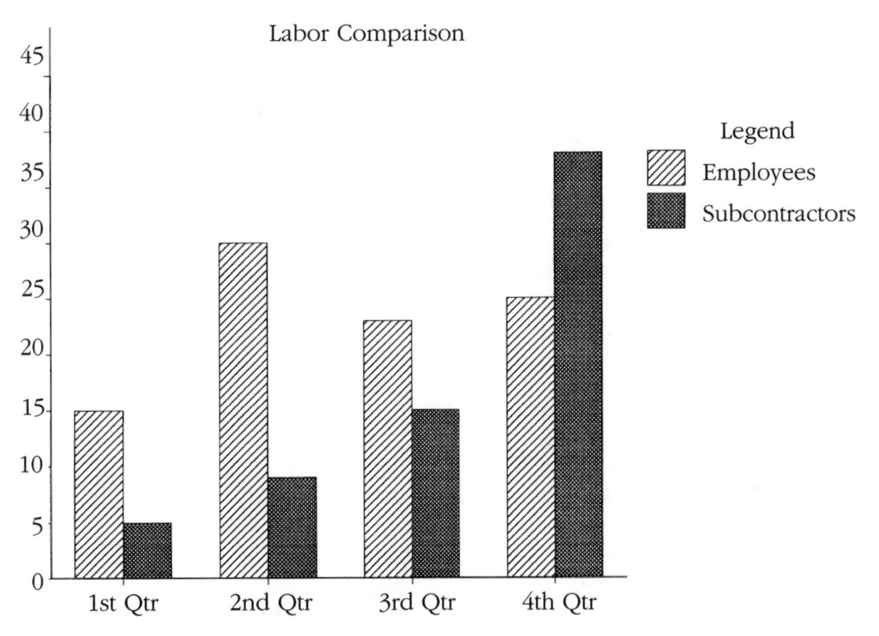

9-9 Labor comparison graph that can be created with a computer.

Inventory control

Inventory control is made faster and more efficient with the use of a computer. As your stock is depleted, the computer can tell you what to order. At the end of the year you won't have to spend hours going into the back room to count your inventory. All you have to do is ask the computer to print a report, and in moments your inventory will be done.

Personal secretary

With a memory-resident program, your computer becomes a personal secretary. The machine will beep to let you know it is time for your next appointment. You can store phone numbers in the computer and have it dial your calls for you via a modem. The right software will all but replace your old appointment book. With the touch of a hot key, you can scan all of your appointments, as far in advance as you like.

The fact is, computers and software can tackle about any job you throw at them. I'm not suggesting that people should be replaced with machines, but there is definitely a place for computers in your business.

BUILD CUSTOMER CREDIBILITY

Computers can lend an air of distinction to your company. In some circumstances, the credibility will be gained as a result of customers having direct contact with the computer. In other cases, the confidence is built by what

Checks and Deposits

		Starting Balance			$1,599.44
323	1/2	Office Rent	$566.81		$1,032.63
324	1/4	Service Station	$ 9.50		$1,023.13
325	1/5	Stationery Store	$ 14.90		$1,008.23
	1/6	Kile Job		$ 100.00	$1,108.23
326	1/10	Electric Company	$112.88		$ 995.35
327	1/10	Natural Gas Company	$ 66.81		$ 928.54
328	1/10	Telephone Company	$ 56.88		$ 871.66
	1/11	Baker Job		$ 345.00	$1,216,66
329	1/14	Plumbing Supplier	$ 99.50		$1,117,16
	1/15	Salary		$1,388.41	$2,505.57
330	1/19	Nails	$ 55.90		$2,449.67
331	1/21	Permits for Dean Job	$123.45		$2,326.22
332	1/22	Tarp	$ 36.22		$2,290,00
333	1/27	County Tax Department	$ 44.90		$2,245.10
	1/30	Salary		$1,388.41	$3,633.51
334	2/2	Office Rent	$566.81		$3,066.70
335	2/4	Hardware Store	$ 44.20		$3,022.50
336	2/5	Electrician-Drill Job	$588.40		$2,434.10
	2/6	Apex Job		$ 345.00	$2,779.10
337	2/9	Electric Company	$105.33		$2,673.77
338	2/9	Natural Gas Company	$ 63.55		$2,610.22
339	2/9	Telephone Company	$ 65.73		$2,544.49
340	2/11	Post Office	$20.00		$2,524.49
341	2/11	Office Supplies	$ 19.44		$2,505.05
	2/15	Salary		$1,388.41	$3,893.46
342	2/20	Shipping Expense	$ 15.00		$3,878.46
343	2/20	Blueprint Copies	$ 40.00		$3,838.46
344	2/23	Gas Station	$ 10.00		$3,828.46
345	2/27	Bill's Restaurant	$ 45.90		$3,782.56
	2/28	Salary		$1,388.41	$5,170.97

9-10 Checks and deposits report.

the computer allows you to do. Let's see how these two different approaches work.

Direct-contact approach

The direct-contact approach is the method most often recognized. At a basic level, your customers come into your office and comment on your computer system. They are impressed by the visual effect of the hardware.

Demonstrating the power of your computerized system is another way to build credibility through direct contact. For example, assume you are a home builder. A young couple comes to your office to discuss house plans.

The couple has a rough sketch of what they want, but it is not drawn to scale. Other builders they have visited have made photocopies of the sketch and let it go at that. Because you have a computer, you are going to make a lasting impression on this young couple.

When the couple produce their well-worn line drawing, you look it over. Then you get up and put it on your flat-bed scanner. The couple watches, at first thinking you are making a normal copy of the drawing. When they see their sketch come up on your computer screen, they are amazed. But the show's not over yet.

Now you take your computer drawing aid and begin drawing lines on the monitor. In less than 15 minutes, you have converted the rough sketch into a scale drawing of the couple's dream home. As you talk to the potential customers, you make adjustments to the on-screen drawing, moving the kitchen sink down the counter, adding an island work space, drawing in a fireplace, and so on. In less than an hour, you have a viable plan drawn for the anxious home buyers.

Now tell me, which contractor would you hire to build your house, the one who made a photocopy or the one that took the time and had the technology to produce a professional drawing? I think the answer is obvious. Even if the computerized contractor wanted more money for the same job, you would probably assume he was worth it. As you can see, with a little thought you can use computers as sales tools.

Indirect approach

The indirect approach relies not on what the system looks like, but what you do with it. Assume for a moment that you are a homeowner. You are soliciting bids for a new family-room addition. During your search, you narrow the field of general contractors down to two. The two contractors are evenly matched in their prices, references, and apparent knowledge of building and remodeling. Who will you choose? You must look deeper to find what separates these two contractors. You start your reconsideration by going back over the bid packages.

The first contractor, Al's Additions Unlimited, has submitted a bid package similar to that of the other general contractors. You received a fill-in-the-blank proposal form, the type available in office supply stores. The proposal had been typed, but it was obvious the preparer was not a typist. Big Al's references checked out, even if they were handwritten on a piece of yellow legal paper. The plans Al submitted were drawn in pencil. The graph paper was stained, probably from coffee. While there was a spec sheet in the package, it was vague. In general, Al's bid package was complete, but it wasn't very professional.

Al's competitor, Renaissance Remodeling, Ltd., submitted essentially the same information, but the method of presentation was considerably different. The contract from Renaissance Remodeling was clean and neatly printed with a laser printer. The paper the contract was printed on was of a heavy, high-quality stock. The accompanying specification sheet was very thorough. It listed every item and described the items in complete detail.

The reference list supplied by Renaissance was printed with the same quality as the contract, and the references were all satisified customers. The plans in this bid package were extraordinary. These plans were drafted on a computer and showed elevations, floor plans, and cross sections. The bid package from Renaissance Remodeling, Ltd. was compiled in an attractive binder. This bid package exuded professionalism. Since all factors, except the bid packages, seem equal, you decide to give the job to Renaissance.

If this were a true story, which contractor would you have chosen? I believe you would have picked Renaissance Remodeling, Ltd. Sometimes all it takes to win the job is a good presentation. Computers can help you make fantastic presentations.

SPREADSHEETS, DATABASES, AND WORD PROCESSING

Spreadsheets, databases, and word-processing software applications are what the majority of business owners use the most. You can buy each of these programs as stand-alone software, or you can buy a software package that incorporates all of the features into one package. Most of these combo-packages are called integrated packages.

Should you buy an integrated package or stand-alone programs? Both types of software can be good, but the answer to the question will depend on your needs and desires. If you are new to computers, an integrated package might be your best choice. Many stand-alone programs are more complex than the individual modules in integrated packages. Let's take a look at some of the pros and cons of each type of software.

You don't have to spend a fortune for software. There are many good deals available on software. Once you get into the computer-scene, you will find a flood of inexpensive software. Some of this bargain-basement software is great, and some of it leaves a lot to be desired. However, for less than $5, you can test drive the software with a demo disk.

By shopping, you can find almost any software combination you want. If you opt for a major-brand software, you may need some time to get acquainted with your purchase. However, if you buy brand-name software, there are some excellent reference books available to help you learn the software applications. If you go for the inexpensive, generic software, you will have more trouble finding aftermarket resource directories.

Spreadsheets

Spreadsheets serve many business functions. Choosing the best spreadsheet for your business needs may require some research. There are dozens of programs that can get the job done. Some of these programs cost hundreds of dollars and others can be had for about $50.

One of the most popular spreadsheet programs has spawned a low-cost, generic competitor. The name-brand program will cost upwards of $250; the clone will run about $50. Other than one function, a financial function that deals with depreciation, I haven't found any substantial difference between the two, other than the price of course.

Most spreadsheets will do approximately the same functions. Some are easier to learn than others, but the end result will be about the same. When you are ready to buy a spreadsheet program, do your homework. There are many good programs available, but there seems to be one leader in the field. I don't want to name names, but if you can count to three, you will be on the track to the best-known and most used spreadsheet on the market.

Databases

Databases are electronic files. These computerized files are generally more efficient than manual files. Think of database software as your opportunity to file and retrieve any information you want. Sorting and retrieving information in a database is easy. Printing reports and mailing labels is no problem when you have a database program. If you keep it in a filing cabinet, you can keep it in a database.

Word processing

Word processing is probably one of the most used types of software available. This type of software is great! You can write and store form letters. You can merge names with the form letters to personalize your mailing. If you make a typing mistake, you don't need correction tape or fluid. With word-processing software you simply type over your mistakes. Word-processing programs can be used for all your correspondence needs.

Word-processing programs don't have to be complicated. You need a word-processing package that will do what you want it to do. This could mean buying a stand-alone program or an integrated system. If you want your words to be perfect, there is one software that assures your success. You will get a spell checker, a thesaurus, graphic capabilities, and much more. This stand-alone, comprehensive software can be purchased for less than $300. Frankly, most first-time users don't need this much power. With the software that makes your words perfect, you can produce newsletters, books, graphics, and just about anything else.

If you don't want to spend the money for a top-notch word-processing package, you can invest in software with less extensive features. Your decision to go with less expensive software shouldn't affect the results of your most basic business needs. It is only when you want to spread your wings that you will need the big-name software.

Integrated programs

Integrated programs can fulfill all of your basic needs. These programs compile spreadsheet, database, and word-processing software into a single package. These combo-packages are efficient and relatively inexpensive. While most integrated programs are not full-featured software, they can satisfy the needs of many small- to average-size businesses.

I have used a few integrated packages. My first integrated package did a great job. Even today, I still use an integrated program for my database needs. Because of my writing requirements I use a stand-alone, major-market word-processing program. My spreadsheet applications are done on a very inexpensive clone of a major spreadsheet program.

If you are new to computer operations and need multiple features, as most businesses do, an integrated software package will serve your needs well. Unless you are willing to spend extensive time learning what you need to know, integrated packages should provide the easiest learning curve. Don't get the wrong idea. Integrated software is not a compromise. It can be an excellent choice.

CHOOSE THE BEST SOFTWARE

Choosing the best software for your business can get confusing. There are hundreds of types of software from which to choose. Many of the various types of software perform about the same functions, with only slight differences. The costs for these software packages can range from less than $20 to well over $700. So how will you ever decide on the right software?

Finding software that does what you want it to do isn't too hard. But finding software that does what you want it to do, in a way that you want it to do it, can be more difficult. When it is time to buy software, do some research.

The first step to acquiring the proper software is to know what tasks you want the program to handle. If you know you want word-processing software that has a built-in spell checker, you know part of what to ask for. However, the list of software packages that have an on-board spell checker is very long. So, you have to narrow your search parameters.

One of the best ways to choose your software is in person. Go to a store that will let you work with the software so you get a feel for the features offered in the package. You can also assess the difficulty of the learning curve for the software. Some software sellers offer demo disks. These demo disks allow you to work with the software on your own computer, at your own speed. Of course the demo disks are not full-featured, but you can gain enough insight to make a buying decision.

Don't buy software until you are sure you want it. Most computer software is not returnable unless it is defective. Since computer disks can be copied, the disks are not allowed to be opened and returned.

If you are new to computers, a good integrated-software package is probably the best place to start. By purchasing a do-it-all package, you can experiment and see what type of work you do the most. If you outgrow the integrated package, you can step up to a stand-alone program.

COMPUTER-AIDED DRAFTING PROGRAMS CAN CHANGE YOUR LIFE

If you deal with floor plans, riser diagrams, wiring diagrams, or any other type of technical drawings, you can benefit from a computer-aided drafting (CAD) program. CAD programs allow everyone to draw professional plans and diagrams. I'm a good example of the type of person that can benefit from a CAD program.

My wife is a fine graphic artist. Kimberley can envision and draw about anything she wants. I, on the other hand, have trouble drawing anything detailed. When I bought my first computer-drafting program I had my doubts. I purchased the software to offset my frustration with freelance artists. You see, I bought the software to illustrate my books. This book doesn't contain the type of illustrations that require a CAD program, but many of my other books do.

My first attempt at CAD was a miserable disappointment, but my further attempts produced excellent results. After I got into computer drafting, I saw a way to improve my contracting business. I gave you an example earlier about how a contractor captures business with a CAD system. This procedure has proved successful for me.

The results you can achieve with a CAD program are unlimited. Once you get the hang of putting drawings on the computer, you are likely to enjoy it. You may even find computer drafting to be therapeutic.

HARDWARE CONSIDERATIONS

While some people buy computer equipment to satisfy their curiosity or their got-to-have-it impulses, selecting the hardware for your computer is serious business. What should you look for in computer hardware? Well, let's find out.

Desktop computers

Desktop computers are far and away the most popular computers among business owners. Desktop units are available in many configurations and speeds. No matter what your needs are, there is a desktop computer available to meet them. Desktop computers can be purchased for less than $1,000, or you can spend upwards of $10,000. How can there be such a disparity in prices? The difference in price is determined by many factors. One factor is name recognition; famous names command more money. Other factors include: speed, memory, features, and more.

Laptop computers

Laptop computers are very popular. These mini-powerhouse computers are capable of the same tasks as many desktop units. However, the compact size and portability of laptops make them especially desirable. For busy people on the move, a laptop is the ideal computer.

You might think you have to give up features for portability, but you don't. It is feasible to get as many features in a laptop as you will find in the average desktop unit. It is not uncommon to find laptops with huge hard drives and blazing clock speeds. Couple these features with a VGA screen and the freedom of battery-operated power, and you've got a winner.

Notebook computers

Notebook computers resemble laptop computers. Notebook computers are typically smaller and lighter than laptops. For this size advantage, you will normally have to sacrifice some features. Notebook computers are fine as a second computer, but I wouldn't recommend one as a primary computer.

Computer memory

How much computer memory do you need? Older computers operated with 512 kilobytes (K) of random-access memory (RAM). There are still many computers available with 512K of memory. This is adequate memory for running most software, but much of the newer software requires more RAM.

If you don't want to be restricted in software options, buy a computer with 640K of memory. Most desktop units and many laptop computers are able to add expansion memory. One megabyte of memory is not uncommon and memory expansion can go much higher.

If you will be doing a lot of CAD work, you will probably want at least two megabytes of memory and possibly more. If your computer doesn't have enough memory, complex drawings will not be able to load from the files into the computer's memory. This means you will not be able to get the drawing on your monitor.

Computer speed

Computer speed is another issue to consider. How fast is fast enough? Some applications will be painfully slow with older, slower computers. For example, if you are doing CAD work a slow computer will drive you crazy. In the time it takes a slow computer to redraw your drafting project, you can get a cup of coffee, open your mail, and pace around the computer. If you plan to use CAD programs, get the fastest machine you can afford.

If word processing will be the primary function of your computer, speed and memory are not as important. Almost any computer will allow you to use word-processing software.

Database work is more enjoyable with a fast computer, but high speed is not essential. Memory requirements will depend on the type of database software you have, but most programs will work with minimal memory.

How do you judge the speed of a computer? The first way is by name. For example, an XT computer is slower than an AT computer. You can get an idea of a computer's speed by its number designation. A 486 machine is extremely fast, a 386 machine is very fast, and a 286 machine is fast.

To determine the exact speed of individual computers, you must look at their megahertz (MH) ratings. The higher the number in the megahertz rating, the faster the machine. For example, not all 386 machines run at the same speed. You might find one 386 with a speed of 16MH and another with a speed of 20MH. The machine with the 20MH rating is the faster of the two.

Hard drives

Hard drives or hard disks are storage devices found inside a computer. Not all computers have a hard disk. Some computers operate on floppy-disk drives. If you are planning to do serious work on your computer, you want a hard drive. Hard drives are faster and more convenient than floppy drives. Another reason to buy a hard drive is the fact that many software programs cannot be used with a floppy-disk drive.

Hard drives come in many sizes. Their size determines how much data can be stored. For example, a 20-megabyte hard drive will hold only half the data that a 40-megabyte drive will. What size hard disk do you need? As a minimum, you should have a 20-megabyte hard drive. A 40-megabyte drive will allow you more room for expansion. If you are planning to store a high volume of graphic files and CAD files, a much larger hard drive will be required.

Floppy-disk drives

Floppy-disk drives can be mounted in the housing of the computer, or they can be independent units that connect to the computer with a cable. Not long ago, many computers depended on floppy-disk drives for their operation. Today hard drives are taking over. Even when computers have hard drives, they usually have floppy drives as well. Common floppy-disk drives come in two sizes: 5¼ inch and 3½ inch. The most popular size is the 3½-inch floppy-disk drive.

CGA monitors

CGA monitors are color monitors. They are at the low-end of the price range for color monitors. If you will only be using your computer for short periods of time, a CGA monitor will do fine. If, however, you will be spending hours staring at the screen, you owe it to your eyes to get a better monitor. Also, if you will be doing graphic work, you will want a more expensive monitor. Before you buy any monitor, make sure it is compatible with your computer. Not all monitors will work with all computers.

EGA monitors

EGA monitors are a step up from CGA monitors. However, if you want more than a CGA monitor, I recommend moving up to a VGA monitor.

VGA monitors

VGA monitors are much easier on your eyes than the less expensive monitors. If you are doing graphic work, you almost have to have a VGA monitor. When you spend your whole day in front of the monitor, a VGA is much easier on your eyes. The cost difference between a CGA monitor and a VGA monitor is not too great—about $200. Once you see the difference, you will get a VGA.

Laser printers

Laser printers are considered the best printers you can buy, and their prices reflect those feelings. Laser printers do turn out beautiful work, and they offer creative options not available with other printers. However, most contracting businesses can get by without a laser printer.

If you plan to do graphics and CAD work, a laser is well worth the expense, but for letters and reports, a dot-matrix printer will get the job done.

Letter-quality printers

Letter-quality printers produce documents that appear to have been typed on a typewriter. These printers do neat text, but they are dreadfully slow.

Dot-matrix printers

Dot-matrix printers can serve most of your needs. These impact printers are fast. When you want to check your job costs, you can run off a report in draft-mode, before you can walk across the office. If you want your text to be dark and pretty, you can switch to the near-letter-quality mode. In this mode the printer is about twice as slow as in draft-mode, but it is still faster than a letter-quality printer. The quality of the text produced with a dot-matrix printer can't compare with that of a laser or letter-quality printer, but it is fine for most business applications.

Other printers

There are, of course, other types of printers. There are thermal printers and printers that spray ink on the paper. With a little shopping, you can see all of the available printers. However, the printers I detailed above are the most popular.

Modems

Modems are devices that allow computers to talk to each other over the telephone lines. Some modems are inside computers and others sit outside the computer and attach with a cable. You can add a modem at any time, and you probably won't need one in the beginning.

Mouse

A mouse, in computer terms, is a device used for drawing on the computer and for clicking on commands. If you will be doing graphic work, you will most likely need a mouse. For everyday office work, you should be able to get by without a mouse.

10 Equipment, vehicles, and inventory

Equipment, vehicles, and inventory will account for most of your start-up costs. If you are already in business, these same items consume a large portion of your operating capital. There is no question that buying an $18,000 truck can put a dent in your bank account. Even with a 10-percent down payment and a 5-year loan at 10-percent interest, you are going to have to shell out $1,800 in cash and about $344 a month for payments. Add to this the cost of registration, taxes, and insurance, and you have a major expense.

Then, depending on the type of business you have, you may have to buy equipment. Hand tools are expensive and power tools are even worse. As for inventory, a service plumber can easily invest $10,000 for in-truck inventory. With these types of expenses, your savings account may need a transfusion.

A large number of businesses are forced to close each year because of the costs associated with equipment, vehicle, and inventory purchases. Some business owners opt to lease vehicles. This saves up-front money and the payments are usually less. But, if the vehicles have excessive mileage or body damage when the lease is up, the end costs may exceed the initial purchase price of the vehicle.

Not only is the cost of acquiring inventory steep, keeping employees from stealing it can be a problem. Somewhere among the maze of options in these three big-ticket items is a happy medium. This chapter is going to help you find your way through the perilous maze to financial safety and success.

LEASING VS. PURCHASING

Both options offer advantages and disadvantages. It makes a lot of sense to rent tools that you only use a few times a year. Leasing vehicles can save you money and provide tax advantages. So, what should you do? Well, let's find out.

Renting tools

Renting expensive tools can be a very smart move for a new business. When you first start out, money can be especially tight and you may not have a real handle on what tools you will need. Renting what you need, when you need it, has several advantages. You can try the tool before you buy it. You can evaluate how badly you need the tool, and you can determine how much extra money you can make with the tool.

When I started my first plumbing business I couldn't afford to buy a large electric drain cleaner. My plumbing business was centered around remodeling and new construction. I didn't have a lot of call for cleaning drains, but I hated referring the calls I did get to other plumbers. I decided to try renting the tool. I rented the drain cleaner for about six months. This gave me time to evaluate how often I used it and how much money I made with it. It became obvious that I could increase my profits by buying such a tool. I bought a good drain cleaner and it paid for itself in a matter of months. After that, all the money I made with it was gravy. It was inconvenient having to run to the rental center to pick up the snake, but the process enabled me to make a wise buying decision. You can do the same type of experiment.

Buying tools

Buying specialty tools can be a mistake. If you buy a bunch of tools you don't use very often, you will deplete your cash for a lost cause. For example, I used to do a high volume of basement bathrooms. Installing these basement baths required the use of a jackhammer. I started out renting a hammer and wound up buying one. For me, this was a good move. However, it might a superfluous expense for some plumbers who rarely need to break up a concrete floor.

Before you buy specialty tools, make sure you need them. The best way to assess your needs for specialty tools is to rent them when you need them and keep track of how they affect your business. If your profits increase, consider purchasing the tool. If you find you only rent the tool a few times a year, continue to use rental tools.

Leasing vehicles

The good points of leasing are:

- minimal out-of-pocket cash,
- normally lower payments,

- possible tax savings, and
- possible short-term commitments.

The disadvantages are:

- no equity gain,
- more concern for the condition of the vehicle, and
- possible cash penalties when the lease expires.

Most auto leases only require the first month's rent and an equal amount for the down payment. In other words, if the vehicle is going to cost you $250 a month, you will need $500 up front. To purchase the same vehicle, you might need $1,000 to $1,500 for a down payment.

The monthly payment on a leased vehicle is usually lower than the payment for a purchased vehicle. How much you save will depend on the cost of the vehicle, but the savings can add up. However, you don't own the vehicle, and at the end of the lease you will have no equity in the truck or car.

Leases usually can be obtained for any term, ranging from one year to five, and sometimes more. Two-year leases are popular, and so are four-year leases. By leasing a vehicle for a short time, like two years, your company fleet can be renewed every two years. This keeps you in new vehicles and presents a good image. However, if the vehicle is not in good shape at the end of the lease, you will pay a price for the abuse.

Most leases allow for a certain number of miles to be put on the vehicle during the term of the lease. If the mileage is higher than the allowance when the lease expires, you will have to pay a price per mile for every mile over the limit. This can get expensive. Additionally, if the vehicle is beat up you can be charged for the loss in the vehicle's value. This can also amount to a substantial sum of money.

Then there are the tax angles. Since I'm not a tax expert, I recommend you talk with an expert. It is likely that leasing will be more beneficial to your tax consequences than buying, but check it out.

Purchasing vehicles

A lot of companies buy their vehicles. When you purchase your vehicles, you build equity in them. If you abuse the vehicle, you will lose money when you sell or trade it, but you won't be penalized by a lease agreement. If your trucks will lead a hard life, purchasing them is probably a good idea.

SEPARATE NEEDS FROM DESIRES

Too many contractors want to have the best and most expensive items available. This is not only unnecessary, it can kill your business. If you can do the job with a $15,000 van, don't buy a $20,000 truck. There is a big difference between a need and a desire. Let's examine how you should decide what to buy.

Let's begin by looking at inventory items. Most new business owners want a well-stocked inventory. They often wind up overstocked. This ties up their cash and sometimes can result in lost money. Before you buy inventory items, figure out what you need, not what you want.

Tools and equipment are areas where many business owners have little willpower. They all seem to want every tool and piece of equipment they could ever hope to use. It's one thing to want it and quite another thing to buy it. It is wise to have the tools and equipment you need, but it is senseless to buy expensive items that you will rarely use. For these occasional-use items, rent them.

The need-and-desire angle comes into play with vehicles too. For years I wanted my own dump truck. I thought it would be great for hauling debris away from my remodeling jobs. I came close to buying one, so close that I was sitting down with the salesman. But at the last minute, I bought a van instead. Why did I buy the van? The van would work for my remodeling work and it would get another crew on the road. In short, the van would pay for itself and the dump truck wouldn't. This is the type of consideration you must give yourself time to make.

FINANCIAL JUSTIFICATION

Just like my dump-truck example, you have to see if what you are about to do is economically feasible. You must prove to yourself the financial benefits of your purchase by calculating the payback period. I gave a brief example earlier about how I rented a drain cleaner before buying one. Let's look at the steps I went through to justify financially that purchase.

To justify the purchase of a power drain cleaner, I rented the tool whenever I had a need for it. I rented the tool for about six months. During this time I kept track of the number of times I used the equipment, what I paid in rental fees, the money and time I lost picking up and returning the unit to the rental center, and how much money I grossed on the use of the snake.

Then I sat down and ran the numbers in order to evaluate my need to purchase this tool. I saw clearly that the power drain cleaner would be a good investment for my company. The new tool, with accessories, cost about $1,200. At that time, the rental fee for the unit was about $20. During normal business hours I was charging $65 for the machine and a mechanic to operate it. My normal hourly fee was $30, so I was essentially charging $35 for the use of the machine. Of this $35 I had to pay the rental store $20. Still, I was getting work I couldn't get without the machine.

The cost of paying a plumber to pick up and return the machine ate up the money I was making off the equipment rental, but I was still making my normal profit from the service call. I deduced that if I owned the machine, I would have no lost time, and my profit per call would increase by $35. Of course, I would have to pay for the machine before I could see this additional profit.

I was averaging six calls a week for the drain cleaning machine. With this volume of activity, it made sense to buy my own drain cleaner. If the demand continued, as it had for the previous six months, the machine would pay for itself in about six weeks. After the machine was paid for, I would enjoy higher net earnings for the remaining life of the machine.

I bought the machine and my plans worked out perfectly. In less than two months the machine was paid for and I continued to see increased revenue from its use. This is a prime example of how you financially justify a purchase.

STOCKING YOUR INVENTORY

The sarcastic answer would be to stock just as much as you need and not a bit more. But the real answer is harder to come by. Inventory requirements vary with different types of businesses. Where a service plumber might need a rolling stock of $7,000, a new-construction plumber could get by with an inventory of less than $2,000. To establish your inventory needs you must assess your customers' buying trends.

Planned jobs

Planned jobs don't require much inventory. You can order a specific amount of material for the job when the job starts. You will want some inventory to make up for any items you forgot in the big order, but on-truck inventory needs for this type of work are minimal.

Impulse buying

If one of your plumbers is at a house fixing the kitchen sink, the homeowner might ask the plumber to install a new lavatory faucet while he's on the job. The customer knows this will save travel time and money. If your plumber's truck has an acceptable lavatory faucet on it, you've made a sale. If the plumber has to go get the faucet, the customer will probably put off the expense and trouble of swapping the faucet. If you will be doing work where the customer is likely to ask you to do more work while you are already on the job, stock some standard impulse items.

Time-savers

A good inventory of frequently used materials can be a real time-saver. If you are on the job and need an extra 50 feet of electrical wire, you will save time and money by having it on the truck. It is a good idea to carry a rolling stock of your most frequently used items, but don't get carried away. Remember, a service truck doesn't need 250 copper ells.

Shop stock

Shop stock is convenient, but it can be a drag on your cash flow. Limit shop stock to what you will use in a two-week period. As you use the material, order new stock. This keeps your inventory fresh and your money turning over.

CONTROL INVENTORY THEFT

Inventory theft is a problem that can hit any company that has employees. Whether you have 1 employee or 100 employees, you could be getting ripped off. Lost inventory is lost money. You must take steps to ensure that your employees are honest and that the material is going where it belongs.

The best way to reduce inventory pilfering is to keep track of your inventory on a daily basis. Have your workers fill out forms for all material used (FIG. 10-1). Have the forms completed and turned in each day. Let employees know, in a nice way, that you check inventory disbursement every week. This tactic alone will greatly reduce the likelihood of your employees stealing from you.

Another way to control the loss of shop stock is to issue all stock yourself. If you don't allow employees access to your inventory, they can't steal it. Rolling stock is harder to track. However, if your employees know you keep daily records of your stock, they will be less likely to empty your truck. It never hurts to do surprise truck inspections and inventories. When you do this, make sure the employees see you inspecting the trucks. The fact that you go on the truck and audit the inventory will reduce your losses.

Truck Inventory

Truck number: ______________________________

Driver: ______________________________

Item	*Size*	*Color*	*Brand*	*Quantity*

Inventory taken on the ______th of ____________________, 19____

10-1 Truck inventory form.

Inventory Control for Trucks

Item	*Quantity*	*Job Name*	*Employee*	*Truck#*	*Date*

10-2 Inventory control form for trucks.

STOCK YOUR TRUCKS EFFICIENTLY

Stocking your trucks efficiently is critical for success and maximum profits. Service vehicles should be set up with everything they need and nothing they don't. I know this is easier said than done, but it's true. Keep your rolling stock to a minimum. You never know when the truck will be broken into or stolen. It is more difficult to monitor employees that deal with mobile inventories. If the inventory on your trucks is collecting dust, get rid of it and don't reorder.

A good way to establish your truck-stock needs is to keep track of the materials you use off the trucks. If you track your material usage, you will stock your trucks with materials you are likely to use (FIG. 10-2). If you have some slow-moving items on the trucks, don't replace them when they are sold. Conversely, if you have some hot items, keep them on the trucks.

Stocking your trucks efficiently will take a little time; start small and work your way up. If you have been in the trade for long, you will have a good idea of your inventory needs. By tracking your material sales, you can perfect your rolling stock.

11

Subcontractors, suppliers, code officers, and materials

If subcontractors do sub-par work, your company's reputation suffers. If suppliers don't maintain delivery schedules, your crews can come to an abrupt halt. Code officers can also make your company look bad. When the work your company produces fails inspection, your business loses credibility. Inferior materials have the power to make your best work look bad. If you want your business to be successful, you must learn how to handle subcontractors, suppliers, code officers, and materials.

IF YOU ARE A CONTRACTOR

If you are a contractor, how your business is perceived will depend upon your skills as a manager. Let's look first at how subcontractors can affect your business.

Subcontractors

As a contractor, it is common to hire subcontractors to perform various types of work. The quality of the work done by these subcontractors is very important. They represent your company. Choosing subcontractors is much like hiring employees. When you put subcontractors in contact with your customers, you are trusting them to maintain the reputation you've worked hard to earn.

Good subcontractors can make you look great. If you have a deep stable of subs, you can respond to work quickly and efficiently. Customers love to get fast service, and subcontractors can give you this desirable dimension. Subcontractors with good work habits and adequate people skills will build your business. While some subcontractors will attempt to steal your customers, most will be happy to maintain their relationship with your company. A lot of subs don't want all the responsibilities that go along with being the general contractor. If you take good care of your subs, most of them will take good care of your business.

Suppliers

You might think that suppliers couldn't have much of an effect on the public's opinion of your business, but they can. As a contractor, you are held responsible for everything that happens on the job. If your supplier's delivery truck damages the customer's lawn, you're going to catch the heat. When materials are not delivered on time, customers are not going to call the suppliers to complain; they are going to call you. As the general contractor you will take all the abuse.

If you want your customers to remain happy (and what contractor doesn't?) you must be in control of the job. This control extends from getting the permit to doing the punch-out work, and everything in between.

The right suppliers can improve your customer relations. If delivery drivers are courteous and professional, the customers will appreciate it. When deliveries are made on time, customers are satisfied. Seeing to it that suppliers make and maintain good customer relations is up to you. You have to lay down the rules for your suppliers to follow. If the suppliers are unwilling to play by your rules, find new suppliers.

Materials

Materials can have a large effect on your customer's peace of mind. If shoddy material shows up on the job, your customer isn't going to be pleased. If the wrong materials are shipped, you will lose time and your customer will lose patience. Don't overlook the important role that materials play in the way customers view your business.

CAREFULLY CHOOSE YOUR PRODUCT LINES

When you choose the proper products, they will sell themselves. As a business owner, you can use all the help you can get, so carry products the public wants. Decide on what products to carry by doing some homework. Read magazines that appeal to the type of people you want as customers. For example, if you want to concentrate on kitchen and bathroom remodeling, read magazines that cater to this type of work. Observe advertisements in the magazines you decide fit your business plan. Pay attention to these ads and you will get a good idea of what your customers want.

Walk through local stores that carry products you will be selling and competing lines. Take notes as you cruise the isles. Pay attention to what is on display and how much is being charged for the items. This type of research will help you target your product lines.

Ride around town to other job sites and see what types of doors, windows, siding, shingles, and similar items are being used. This on-site investigating will put you in touch with what the public wants.

The most direct way to determine what customers want is to ask them. Go door to door and do a cold-call canvassing of a neighborhood. You will experience a lot of rejection, but you will also get some answers. If you don't like knocking on doors, use a telephone. You can even have a computer make the phone calls and ask the questions for you.

If you don't want to use face-to-face techniques or telephones, use direct mail. Direct mail is easy to target, and it is fast and effective. While mailing costs can get steep, the results might overcome the costs. Design a questionnaire and mail it to potential customers. Done properly, your mailing will look like you have a sincere interest in what individuals want. It will appear this way because you have a sincere interest.

The responses to your questionnaire will tell you what products to carry. You can improve the odds of having the pieces returned by self-addressing the response card. You should also pick up the tab on the return postage. This can be done by purchasing a permit from the local post office and having it printed on your cards, or you can affix postage stamps to the cards, but this will cost more. With the permit from the post office, you pay only for the cards that are returned, not counting the permit fee. If you use postage stamps, you will pay for postage that may never be used.

To convince people to fill out your questionnaire, provide an incentive. One idea for an incentive is a discount off your normal fees. This idea may work, but it will look very commercial. If you design the piece to look like a respectable research effort, more people will respond to your questions.

AVOID DELAYS IN MATERIAL DELIVERIES

If your materials are delayed, your jobs will be delayed. If your jobs are delayed, your payments will be delayed. If your payments are delayed, your cash flow and credit history can falter. If your cash flow dries up, your business is in trouble. Extended delays can result in the loss of your good credit rating or even your business. You can eliminate some of these potentially dangerous situations by avoiding delays in your material deliveries.

Your responsibility for maintaining the delivery schedule begins when you order your materials. There are some basic principles you can apply to keep your deliveries on schedule. Start by getting the name of the person who takes your order (FIG. 11-1). Ask the order taker to document the delivery in writing. While you are at it, get the name of the store manager. You will probably need it.

Once you have the initial delivery date, stay on top of the delivery. If you have placed the order several days in advance, make follow-up phone calls to check the status of your material. Always get the names of the people you talk to. You never know when you will have to lodge a complaint.

If you maintain a presence on the phone or in person, the employees who handle your deliveries will not forget you. They will assume that if you are this attentive now, you will be horrible to deal with if they mess up the order. This intimidation works in your favor.

With a lot of effort and a little luck, your deliveries will be made on schedule. If the shipment does go astray, contact the store manager. Advise the manager of the problem and the ripple-effect it is creating for your busi-

Material Order Log

Supplier: ______________________________

Date order was placed: ______________________________

Time order was placed: ______________________________

Name of person taking order: ______________________________

Promised delivery date: ______________________________

Order number: ______________________________

Quoted price: ______________________________

Date of follow-up call: ______________________________

Manager's name: ______________________________

Time of call to manager: ______________________________

Manager confirmed delivery date: ______________________________

Manager confirmed price: ______________________________

Notes and Comments

____

____

____

____

____

____

____

____

____

____

____

____

11-1 Material order log.

ness. Produce your documentation on the order. Show the manager your written delivery date, employee names, and supporting documentation. This tactic will set you apart from the customers who complain incoherently. You will come across as a serious professional. If you don't get satisfactory results from the manager, move up the ladder to higher management. If you have created a strong paper trail, you will get results.

CHOOSING SUBCONTRACTORS

Choosing subcontractors requires extensive time and effort. You must give this part of your job the attention it deserves. There are some rules you should follow.

Initial contact

Most people act, at least partially, on gut instinct. When you meet subcontractors for the first time, you will develop an opinion. With the potential importance of this meeting, you will want to control the circumstances.

Subcontractors and general contractors are meant to go together, like peanut butter and jelly. If there is not a comfort level between the two parties, the work resulting from the business marriage will not be at its best. If you project a professional image, subcontractors will seek you out. Once subs find you, maintain the business image. Whether you are talking on the phone or in person, send the right messages. Let subcontractors know you are a professional and will accept nothing less from them.

Application forms

While subs are not going to be traditional employees, it is not unreasonable to ask them to complete an employment application (FIG. 11-2). The application you use may not resemble those you use for employees, but you want to know as much about your subcontractors as possible.

The application may contain questions pertinent to the types of work the subcontractor is equipped to do. Asking for credit and work references is a reasonable request. Having the subcontractors list their insurance coverage also is beneficial. You can customize your applications to suit your needs. It may be wise to discuss the form and content of your subcontractor applications with an attorney. You don't want to be guilty of asking inappropriate questions.

Basic interviews

When you conduct your interviews with subcontractors, you will want to derive as much insight as possible into the qualities of the subcontractors. These interviews will be the basis for your decision to use or eliminate subcontractors.

During the interview, control the conversation. Let the subcontractor talk, but don't let the sub run the show. You should set the pace for the in-

Subcontractor Questionnaire

Company name____

Physical company address____

Company mailing address____

Company phone number____

After-hours phone number____

Company president/owner____

President/owner address____

President/owner phone number____

How long has company been in business____

Name of insurance company____

Insurance company phone number____

Does company have liability insurance____

Amount of liability insurance coverage____

Does company have Worker's comp. insurance____

Type of work company is licensed to do____

List business or other license numbers____

Where are licenses held____

If applicable, are all workers licensed____

Are there any lawsuits pending against the company____

Has the company ever been sued____

Does the company use subcontractors____

Is the company bonded____

Who is the company bonded with____

Has the company had complaints filed against it____

Are there any judgments against the company____

Notes and Comments

____

____

____

____

____

____

____

11-2 Subcontractor questionnaire.

terview. If subcontractors run over you in the interviews, they will run over you in the normal course of business.

If you have a professional office, your office is a fine place to meet subcontractors. If your office conditions don't reflect the image you want to give, pick a meeting place that will allow you to project your best image.

Check references

Check the references of subcontractors before you use their services. If a subcontractor has been in business long, there should be references available. Ask for these references, and follow up on them. If you don't confirm the qualities of subcontractors by checking references, you may not be getting the service you want.

Check credit

Another part of the screening process for subcontractors is to check their credit. The credit ratings of subcontractors will tell you much about the individuals and their businesses. If subcontractors have bad credit, it doesn't mean they are bad subcontractors or poor workers. However, if their companies are in trouble, you probably don't want to trust your business to them.

In all of your business endeavors you must learn to read between the lines. Credit reports are a good example where the facts may not tell the whole story. Let's say you are reviewing a credit report and see that a subcontractor has filed for bankruptcy. Would you subcontract work to this individual? If you wouldn't, you may be missing out on a good worker.

The fact that someone has filed for bankruptcy is not enough to rule out doing business with the individual. Individuals can get into financial trouble, without fault of their own. You must be willing and able to decipher what you are seeing. When you learn to read between the lines, you will be a more effective business person.

Set guidelines

If you plan to utilize the services of the subcontractors you are interviewing, set the guidelines for doing business with your firm. If you require all of your subcontractors to carry pagers, let this fact be known in the interview. If your rules require subcontractors to return your phone calls within an hour, make the point clearly. Remember, you are in control, but you can't expect people to read your mind. You have to let your desires be known.

Come to terms

What you want and what the subcontractors want may not be the same. If you are going to do business with subcontractors, you should work out the terms of your working arrangements in advance.

Discuss contracts

Go over your subcontract agreements with the subcontractors. If there are any questions or hesitations, resolve them at the meeting. You don't want to get in the middle of a job and find out that your subs will not play by the rules. The more detail you go into in the early stages of your relationships, the more likely you are to develop good working conditions with your subcontractors. Take as much time as necessary to remove any doubts about the meanings of your contracts.

Maintain the relationship

When you find good subs, you have probably invested a significant amount of time in your acquisitions. To avoid having this time investment wasted, you must work to keep the relationship on the friendly side of the scale. This doesn't mean you have to become buddies with your subs, but you do have to fulfill your commitments.

If you tell a subcontractor that you will pay bills within five days of receipt, you had better be prepared to pay the bills. Stick to the terms in your subcontractor agreements. If you breach your agreements with subcontractors, you will have a great deal of difficulty in getting and keeping good help.

RATING SUBCONTRACTORS

Rating subcontractors may take a little extra effort, but it will be worth it. The rating procedure starts in the interview, but goes deeper (FIG. 11-3). Let's take a look at some of the factors you should consider.

Work history

One of the first qualities you should evaluate is the work history of the subcontractor. When it comes times to hire subcontractors, experience counts. It may not be important if the sub has just started in business, but it is meaningful for the individual to have work experience. For example, a person with 15 years of experience that has just gone into business may be a better sub than the one that has been in business for 2 years, but only has 5 years of experience.

If the subcontractor has been in business for awhile, there is a better chance that his business will last. New businesses often fail, but businesses that have been around for between three and five years have a better chance of survival. Business owners that survive these early years have business experience and dedication. Both are admirable traits in a subcontractor.

Business procedures

One of the most critical aspects of subcontractors is how easy they are to reach by phone. If you can't communicate with your subs, you will have problems. Most subs will tell you how attentive they are, but you should verify their claims. After the contractors leave your office, call them. You know they have

Contractor Rating Sheet

Category	*Contractor 1*	*Contractor 2*	*Contractor 3*
Contractor name			
Returns calls			
Licensed			
Insured			
Bonded			
References			
Price			
Experience			
Years in business			
Work quality			
Availability			
Deposit required			
Detailed quote			
Personality			
Punctual			
Gut reaction			

Notes

11-3 Contractor rating sheet.

just left the meeting and are not in their offices. You will find out how their phones are answered and how quickly they will return your call.

Before you commit to using a subcontractor, conduct a test. Call three subs and schedule each of them for a walk-through inspection of a job. It doesn't matter if the job is real or a dummy. What matters is that the subcontractors believe they are inspecting and bidding a real job.

Why should you play this hoax on innocent contractors? You should do it to see how the contractors respond. Will they be punctual? How long will it take to get the quotes you want? How will the subcontractors behave around the stand-in customer? All of these are conditions that you should investigate before you hire the subs for a real job. If they fail your test, you may have saved a job.

Tools and equipment

If your subcontractors don't have the necessary tools and equipment, they will not be able to give you the service you desire. Don't hesitate to inquire about the tools and equipment the subcontractors possess.

Insurance coverage

Insurance coverage is a big deal. You cannot afford to use subcontractors that are not properly insured. It is easy to lose your business to a court decision. For your own protection, you must make sure your subcontractors carry all the insurance they need.

Liability insurance should be a mandatory requirement. If the sub has employees, other than close family members, worker's compensation insurance is needed. Even if the subcontractor is not required to carry worker's comp, you should have a waiver signed by the business owner. The waiver, which should be prepared by your lawyer, will protect you from claims and insurance audits.

If you use the services of subcontractors that are not properly insured, you may have to pay up at the end of the year. When your insurance company audits you, as they usually do, you will be responsible for paying penalties if you used improperly insured contractors. These penalties can amount to a substantial sum of money. To avoid losing money, make sure your subcontractors are currently insured for all necessary purposes.

Specialties

When you deal with specialists you may pay extra, but the end result may be a bargain. How can you pay more and come out of the job with more profits? While specialists may charge higher fees, they are often worth the extra cost. These subcontractors often do a better job and do it faster. Remember, time is money. When you save time you have a chance to make more money. With this in mind, ask potential subcontractors about their specialties. You may find it cost-effective to use different subs for different jobs.

Licenses

Licenses are another issue you should investigate when you rate subcontractors. If subcontractors are not licensed legally, you can get into deep trouble. It is imperative for you to engage only subcontractors that meet standard licensing requirements. If you use unlicensed subcontractors, you are flirting with disaster.

Work force

A subcontractor's work force is another consideration. You don't want to take on a sub that cannot handle your workload. For this reason, you must know the capabilities of the subcontractors. If you give a small contractor too much work, he may frantically add employees to keep your business, possibly selecting some undesirable help. This undesirable help will make your company look bad. It is better to have multiple subcontractors than to have one contractor that can't give the service you need.

While it is more convenient to work with a single subcontractor, it may not be feasible. There will be times when you will need more than one sub in each trade. As a safety precaution, you should have at least three subcontractors in each trade. This depth of subcontractors will give you more control.

CONTROL YOUR SUBCONTRACTORS

Subcontractors can take advantage of you, if you let them. However, if you establish and implement a strong subcontractor policy, you should be able to handle your subs. It is imperative that you remain in control. If subcontractors have the lead role, your company will be run by the subs, not yourself.

Controlling subcontractors will be much easier when you follow some simple rules. The most important rule is to document your dealings in writing. Other rules include:

- Create and use a subcontractor policy.
- Be professional and expect professionalism from your subs.
- Use written contracts with all of your subcontractors (FIG. 11-4).
- Use change orders for all deviations in your agreement.
- Dictate start and finish dates in your agreement (FIG. 11-5).
- Penalize subcontractors for finishing jobs late.
- Always have subs sign lien waivers when they are paid (FIG. 11-6).
- Keep certificates of insurance on file for each sub.
- Don't allow extras, unless they are agreed to in writing.
- Don't give advance contract deposits.
- Don't pay for work that hasn't been inspected.
- Use written instruments for all your business dealings.

Subcontract Agreement

This agreement, made this ____th day of ____________, 19___, shall set forth the whole agreement, in its entirety, between Contractor and Subcontractor.
Contractor: ____, referred to herein as Contractor.

Job location: ____

Subcontractor: _____________________________________, referred to herein as Subcontractor.
The Contractor and Subcontractor agree to the following:

Scope of Work

Subcontractor shall perform all work as described below and provide all material to complete the work described below.

Subcontractor shall supply all labor and material to complete the work according to the attached plans and specifications. These attached plans and specifications have been initialed and signed by all parties. The work shall include, but is not limited to, the following:

____

____

____

____

____

____

____

____

Commencement and Completion Schedule

The work described above shall be started within three days of verbal notice from Contractor, the projected start date is ______________________. The Subcontractor shall complete the above work in a professional and expedient manner by no later than _____ days from the start date. Time is of the essence in this contract. No extension of time will be valid without the Contractor's written consent. If Subcontractor does not complete the work in the time allowed, and if the lack of completion is not caused by the Contractor, the Subcontractor will be charged fifty dollars ($50.00) per day, for every day work extends beyond the completion date. This charge will be deducted from any payments due to the Subcontractor for work performed.

Page 1 of 3 initials___

11-4a Subcontract agreement (page 1).

Contract Sum

The Contractor shall pay the Subcontractor for the performance of completed work subject to additions and deductions as authorized by this agreement or attached addendum. The contract sum is ________________________________($___________).

Progress Payments

The Contractor shall pay the Subcontractor installments as detailed below, once an acceptable insurance certificate has been filed by the Subcontractor with the Contractor:
Contractor shall pay the Subcontractor as described:

____

____

____

____

____

____

____

____

____

____

____

____

____

____

____

____

____

All payments are subject to a site inspection and approval of work by the Contractor. Before final payment, the Subcontractor shall submit satisfactory evidence to the Contractor that no lien risk exists on the subject property.

Page 2 of 3 initials___

11-4b Subcontract agreement (page 2).

Working Conditions

Working hours will be 8:00 A.M. through 4:30 P.M., Monday through Friday. Subcontractor is required to clean his work debris from the job site on a daily basis and leave the site in a clean and neat condition. Subcontractor shall be responsible for removal and disposal of all debris related to his job description.

Contract Assignment

Subcontractor shall not assign this contract or further subcontract the whole of this subcontract, without the written consent of the Contractor.

Laws, Permits, Fees, and Notices

Subcontractor shall be responsible for all required laws, permits, fees, or notices, required to perform the work stated herein.

Work of Others

Subcontractor shall be responsible for any damage caused to existing conditions or other contractor's work. This damage will be repaired, and the Subcontractor charged for the expense and supervision of this work. The Subcontractor shall have the opportunity to quote a price for said repairs, but the Contractor is under no obligation to engage the Subcontractor to make said repairs. If a different subcontractor repairs the damage, the Subcontractor may be back-charged for the cost of the repairs. Any repair costs will be deducted from any payments due to the Subcontractor. If no payments are due the Subcontractor, the Subcontractor shall pay the invoiced amount within 10 days.

Warranty

Subcontractor warrants to the Contractor, all work and materials for one year from the final day of work performed.

Indemnification

To the fullest extent allowed by law, the Subcontractor shall indemnify and hold harmless the Owner, the Contractor, and all of their agents and employees from and against all claims, damages, losses and expenses.

This agreement, entered into on ______________________________, 19______, shall constitute the whole agreement between Contractor and Subcontractor.

______________________________________ ______________________________________

Contractor Date Subcontractor Date

11-4c Subcontract agreement (page 3).

Commencement and Completion Schedule

The work described above shall be started within 3 days of verbal notice from the customer, the projected start date is ________________________. The subcontractor shall complete the above work in a professional and expedient manner by no later than twenty (20) days from the start date.

Time is of the essence in this subcontract. No extension of time will be valid without the general contractor's written consent. If subcontractor does not complete the work in the time allowed and if the lack of completion is not caused by the general contractor, the subcontractor will be charged one-hundred dollars ($100.00) for every day work is not finished after the completion date. This charge will be deducted from any payments due to the subcontractor for work performed.

11-5 Sample completion clause.

Subcontractor Liability for Damages

Subcontractor shall be responsible for any damage caused to existing conditions. This shall include new work performed on the project by other contractors. If the subcontractor damages existing conditions or work performed by other contractors, said subcontractor shall be responsible for the repair of said damages. These repairs may be made by the subcontractor responsible for the damages or another contractor, at the discretion of the general contractor.

If a different contractor repairs the damage, the subcontractor causing the damage may be back-charged for the cost of the repairs. These charges may be deducted from any monies owed to the damaging subcontractor, by the general contractor. The choice for a contractor to repair the damages shall be at the sole discretion of the general contractor.

If no money is owed to the damaging subcontractor, said contractor shall pay the invoiced amount, to the general contractor, within seven business days. If prompt payment is not made, the general contractor may exercise all legal means to collect the requested monies.

The damaging subcontractor shall have no rights to lien the property where work is done for money retained to cover the repair of damages caused by the subcontractor. The general contractor may have the repairs made to his satisfaction.

The damaging subcontractor shall have the opportunity to quote a price for the repairs. The general contractor is under no obligation to engage the damaging subcontractor to make the repairs.

11-6 Sample damage clause.

DEALING WITH GENERAL CONTRACTORS AS A SUBCONTRACTOR

Most general contractors have good intentions, but it is not uncommon for generals to be slow to pay your bill. Your best line of defense with subcontracting work is a good contract. The contract should state clearly the terms and conditions of all your work and fees.

Most general contractors are reluctant to give contract deposits. Part of their reasoning is cash flow. If a contractor gives each subcontractor front-money, the contractor will have less cash with which to work. A big reason general contractors refuse to give deposits is the potential for loss. The general who gives a generous deposit may never see it or the subcontractor again.

Control is another factor. Giving people money for work that hasn't been done is a sure way to lose control. Money is an excellent lever for keeping subcontractors in line. If a deposit is given, the general has less control.

So, here's your problem. The generals won't give deposits, and you won't work without a deposit. Most subcontractors give in and do the work, hoping to get paid later. The fact is, if you want to work for general contractors you will generally have to work without deposits. But don't do so blindly. Just as a general may lose a deposit, you could lose your labor and materials.

Before working for new contractors, check them out. Get the names of the other subs working for the general. Call those subcontractors and see how they feel about working for the contractor. If you are a member of a credit reporting bureau, get permission to pull a credit report on the contractor. Check with local agencies for any complaints that may have been filed against the contractor. Get a physical address for where the general contractor can be found. If you must have legal papers served on the contractor, it is hard to serve them to a post office box.

Once you have done your homework, proceed with caution. Your contract should stipulate when you will be paid and what your remedies are if the contractor doesn't pay. Check with your attorney on lien rights and legal actions that may help recover your money. By knowing in advance how to deal with deadbeats, you save valuable time and make the right moves when it counts.

Don't become too dependent on one or two general contractors. It is nice to get steady business from the same source, but if something happens to the general you will be out of work. It is wise to spread your work out so that you can survive, even if your best accounts dry up.

Never get too comfortable with your general contractors. It is easy to slack off on paperwork and rules once you get to know someone, but it is a mistake. Keep your relationships on a business level. If you mix too much business with pleasure, you can wind up in a mess.

DEALING WITH SUPPLIERS

Dealing with suppliers is not as simple as placing an order and waiting. It is wise to establish a routine with your suppliers. If you are going to use purchase orders, use them with every order. When you want job names written on your receipts, insist that they are always included. If you allow employees to make purchases on your credit account, set limits on how much each can purchase. Then make sure everyone at the supply house knows which employees are authorized to charge on your account and their charge limits.

Get to know the manager of the supply house. Without a doubt, there will come a time when you and the manager will have a problem to solve. At these times it helps to know each other.

When you begin to use a new supplier, make sure you understand the house rules. What is the return policy? Will you be charged a restocking fee? Will you get a discount if you pay your bill early? What is your discount percentage? Will the discount remain the same, regardless of the volume you purchase? These are just some of the questions for which you should have answers.

CUT YOUR BEST DEAL

Price, service, and quality are the critical factors that must be pondered when you look for the best deal. Getting the lowest price doesn't always mean you are getting the best deal. If you don't get quality and service to go along with a fair price, you are probably asking for trouble. Let me give you an example.

Assume you have ordered roof trusses from your supplier. After shopping prices, you decided to go with the lowest price, even though you have never dealt with the supplier before. The trusses are ordered and you are given a delivery date. All of your work is scheduled around the delivery of the trusses.

The trusses will be used to replace a rotted roof structure. You can't tear off the old roof until you know the trusses are available. On the day of delivery, you call the supplier and inquire about the status of the trusses. You're told the trusses are on a delivery truck and will be at your job by mid morning.

Your crew begins to demolish the existing roof structure. The plan is to have the old roof off and the new roof on before dark. By lunchtime, your crew is at a standstill. The old roof is off, but the trusses haven't arrived.

A phone call to the supplier reveals that the delivery truck broke down on the way to the job. You're told the trusses won't arrive until the next morning. Now what are you going to do? You've got a house with no roof and no way to get the trusses to the job. You should have waited until the trusses were on the job to tear off the old roof, but you didn't. Needless to say, you've got a problem.

When you ask the supplier to transfer the trusses to a different delivery truck, so you can get them immediately, you're told that the supplier doesn't have another truck capable of transporting the trusses. As it turns out, you have to cover the house in plastic and put the residents up in a motel. You have lost time, money, and credibility.

Would this have happened if you had used your regular supplier? Probably not because your regular supplier has enough trucks to make a switch, if necessary. Your great deal on inexpensive trusses has turned into a major flop. So, you see, price isn't everything.

EXPEDITE MATERIALS

Learning the secrets of expediting materials will keep your business running on the fast track. Large companies have people that do little more than expedite materials. For these companies, expediting materials is a full-time job. However, most contractors don't have an employee with the sole responsibility of getting materials to the job. These contractors must do their own material acquisitions and handling. When a person has to tend to multiple tasks, it is easy for some part of the job to be neglected. In contracting, the expediting of materials is often pushed aside to make room for more pressing duties.

All too many contractors call in a material order and forget about it. They don't make follow-up calls to check the status of the material. It is not until the material doesn't show up that these contractors take action. By then, time and money is lost.

A large number of contractors never inventory materials when they are delivered. If 100 sheets of plywood were ordered, they assume they got 100 sheets of plywood. Unfortunately, mistakes are frequently made with material deliveries. Quantities are not accurate. Errors are made in the types of materials shipped. All of these problems add up to more lost time and money.

If you want to make your jobs run smoother, take some time to perfect your control over materials. When you place an order, have the order taker read the order back to you. Listen closely for mistakes. Call in advance to confirm delivery dates. If a supplier has forgotten to put you on the schedule, your phone call will correct the error before it becomes a problem.

When materials arrive, check the delivery for accuracy. Ideally, this should be done while the delivery driver is present. If you discover a fault with your order, call the supplier immediately. You can reduce your losses by catching blunders early.

Keeping a log of material orders and delivery dates is one way of staying on top of your materials. One glance at the log will let you know the status of your orders. When you talk to various salespeople, record their names in your log. If there is a problem, it always helps to know who you talked to last. Get a handle on your materials, and you will enjoy a more prosperous business.

AVOID COMMON PROBLEMS

The two biggest reasons for problems between contractors and suppliers or subcontractors are poor communication and money. Money is usually the largest cause of disputes, and communication breakdowns cause the most confusion. If you can conquer these barriers, your business will be more enjoyable and profitable.

There are few excuses for problems in communication if you always use written agreements. When you give a subcontractor a spec sheet that calls for a specific make, model, color, and whatever, you eliminate confusion. If the subcontractor doesn't follow the written guidelines, an argument may ensue, but you will be the victor.

As for money, written documents can solve most of the problems caused by cash. When you have a written agreement that details a payment schedule, there is little room for argument. Use written agreements to eliminate most of the causes for aggravation and arguments.

BUILD GOOD RELATIONS WITH CODE OFFICERS

If your business depends on the approval of code officers, you will do well to get to know the inspectors. Code officers are often scorned. Contractors that have problems with inspectors cuss them and buck against the system. If these contractors would direct the same amount of energy in a more productive direction, they could solve their problems.

Like it or not, code officers are a fact of life for most contractors. If you want to make your life easier, get to know your code officers. I'm not saying you have to become best buddies, but at least be civil. Your attitude will have a great deal of influence on the posture assumed by the code officer. Don't be afraid to smile and talk with your inspectors. By getting to know each other, problems will be easier to resolve.

AVOID REJECTED CODE-ENFORCEMENT INSPECTIONS

Work is rejected because it is not in compliance with local code requirements. If you know and understand the code requirements, you shouldn't get many rejection slips (FIG. 11-7). If you don't understand a portion of the code, consult with a code officer. It is part of an inspector's job to explain the code to you.

Again, attitude can have a bearing on the number of rejections you get. If you walk around with a chip on your shoulder, inspectors may look a little more closely for minute code infractions. If you play by the rules, you won't have much trouble with the officials. But, don't ever try to put one over on a code officer. If you get caught, your business life will be miserable for a long time to come. Inspectors can be a close-knit group. When you con one, others will get the word and your work will be put under a microscope.

Code Violation Notification

Contractor: ______________________________

Contractor's address: ______________________________

City/state/zip: ______________________________

Phone number: ______________________________

Job location: ______________________________

Date: ______________________________

Type of work: ______________________________

Subcontractor: ______________________________

Address: ______________________________

Official Notification of Code Violations

On March 22, 1993, I was notified by the local code enforcement officer of code violations in the work performed by your company. The violations must be corrected within two business days, as per our contract dated March 1, 1993. Please contact the codes officer for a detailed explanation of the violations and required corrections. If the violations are not corrected within the allotted time, you may be penalized, as per our contract, for your actions in delaying the completion of this project. Thank you for your prompt attention to this matter.

____

General Contractor Date

11-7 Code violation notification form.

12
Pricing your services and materials

Effective pricing of services and materials is an essential element of a profitable business plan. If your prices are too low, you may be very busy, but your profits will suffer. If your prices are too high, you will be sitting around, staring at the ceiling, and hoping for the phone to ring. Somewhere between too low and too high is the optimum price for your products and services. The trick is finding out what those prices are. This chapter is going to help you establish the right price for your services and materials.

MAKE YOUR PRICES ATTRACTIVE

You must establish your pricing structure with research—lots of research. What kind of research is necessary to pick the proper pricing? You can start by calling your competition. For example, if you are opening an electrical business for service and repairs, call around and see what other electricians are charging for their hourly rate. This is simple research. All you have to do is go through the phone book and call the other contractors. When they answer, ask what their hourly rates are. In less than an hour you will know what the majority of your competitors charge for their services. Once you know what the rest of the pack is charging, you can nail down the right price for your hourly fees.

Some newcomers to the contracting business make a serious mistake. When these people learn what their competitors are charging, the rookies price their services far below the crowd in the hope of grabbing all the business. However, if you set your prices far below your competitors, you will alienate yourself from the competition and may lose the respect of potential customers. If your prices are too low, customers may be afraid to use your services. As a new company with low overhead, it's fine to work for less than the well established companies, but don't price yourself too far below your competition.

When it comes to setting attractive prices, look below the surface. There are many factors that control what you are able to earn. Let's take a look at what is considered a profitable markup.

Materials

It is not difficult to project a reasonable markup, but defining a profitable markup is not as easy. Some contractors feel a 10-percent markup is adequate. Others try to tack 35 percent onto the price of their materials. Which group is right? Well, you can't make that decision with the limited information I have given you. The contractors that charge a 10-percent markup may be doing fine, especially if they deal in big jobs and large amounts of materials. The 35-percent group may be justified in their markup, especially if they are selling small quantities of lower-priced materials.

Markup is a relative concept. Ten percent of $100,000 is much more than 35 percent of $100. For this reason, you cannot blindly pick a percentage of markup to be your firm figure. The figure will need to be adjusted to meet changing market conditions and individual job requirements. You can, however, pick percentage numbers for most of your average sales.

If you are in the repair business and are typically selling materials that cost around $20 dollars, a 35-percent markup is fine, if the market will bear it. If you are building and selling houses, a 10-percent markup on materials should be sufficient. To some extent, you have to test the market conditions to determine what price consumers are willing to pay for your materials.

If you are selling common items that anyone can go to the local hardware or building supply store and price out, you must be careful not to inflate your prices too much. Customers expect you to mark up your materials, but they don't want to be gouged. If you install light bulbs in my new light fixture and charge me twice as much for the bulbs as what I could have bought them for in the store, I'm not going to be happy. Even though the amount of money involved in the light-bulb transaction is puny, the principle of being charged double for a common item still exists.

If you typically install specialty items, you can increase your markup. People will not be as irritated paying a marked-up, but reasonable price for an unusual product. A markup of 20 percent will almost always be acceptable on average residential jobs. When you decide to go above the 20-percent point, do so slowly and test the response of your customers.

Labor

This is a question almost every contractor considers. Low prices can keep companies busy, but that doesn't mean the low-priced companies are making a profit. Gross sales are important, but net profits are what business is all about. If a company is not making a profit, there is not much sense in operating the business.

Companies that work with low prices fall into several categories. Some companies work on a volume principle. By doing a high volume, the company can make less money on each job and still make a profit by doing many jobs. This type of company is hard to beat.

Some companies sell at low prices out of ignorance. Many small business owners are not aware of the overhead expenses involved with being in business. For example, a plumber that is making $16 an hour at a job might think that going into business independently and charging $25 an hour would be great. The plumber might even start into business charging only $20 an hour. From the plumber's perspective, he is making at least $4 an hour more than he was at his job. Is he really making $4 an hour more? Yes and no; he is being paid an extra $4 an hour, but he is not going to get to keep much of it.

When this low-priced plumber comes into the business world, he may take a lot of work away from established contractors. Experienced contractors know they can't make ends meet by charging such small fees. But the plumber, an inexperienced business owner, hasn't yet tallied all of his expenses. Once the overhead expenses start eating away at what the plumber thought was a great profit, the per-hour rate may not look so good.

This businessman will soon learn that overhead expenses are a force to be reckoned with. Insurance, advertising, callbacks, self-employment taxes, and a mass of other hidden expenses will erode any profits the plumber thought he was making.

This type of inexperienced businessperson will either go out of business quickly or adjust his prices for services and materials. For established contractors competing against newcomers, being able to endure the momentary drop in sales will be enough to weather the storm. In a few months the new business will either be gone or its services will be priced in a reasonably competitive range.

PRICE YOUR SERVICES FOR LONGEVITY

The price of your services and materials relates directly to your success and longevity. It can be very difficult to decide on the right price for your time and materials. There are books that give formulas and theories about how to set your prices, but these guides are not always right. Every town and every business in town will influence the prices the public is willing to pay. You can use many methods to find the best fees for your business to charge. Let's look at some of them.

Pricing guides

Pricing guides can be a big help to the business owner who has little knowledge of how to establish the value of labor and materials. These guides also can cause you frustration and lost business. Most estimating guides provide a formula for adjusting the recommended prices for various regions. For example, a two-car garage that is worth $7,500 in Maine may be worth $10,000 in Virginia. The formulas used to make this type of adjustment are usually a number multiplied against the recommended price. By using the multiplication factor, a price can be derived for services and materials in any major city.

The idea behind these estimating guides is a good one, but there are flaws in the system. I have read and used many of these pricing books. From my personal experience, the books have not been accurate for the type of work I did. I don't say this to mean the books are no good. There are many times when the books are accurate, but I have never been comfortable depending solely on a mass-produced pricing guide.

Instead, I have found estimating books to be very helpful as a piece of the pricing puzzle. I use them to compare my figures and to ensure that I haven't forgotten items or phases of work. Most bookstores carry some form of estimating guides in their inventory.

Research

Research is one of the most effective ways to determine your pricing. When you look back at historical data, you can find many answers to your questions. You can see how the economy swings up and down. You can see how prices have fluctuated over the years, and you can start to project the curve of the future. Historical data can be found by reading old newspaper ads, researching tax evaluations on homes, and by talking with real estate appraisers.

Real estate appraisers are an excellent source for pricing information. Most appraisers are willing to consult with clients on an hourly basis. By spending a little money to talk with an appraiser, you can save thousands of dollars in lost income. If your business involves providing goods and services for homes and businesses, appraisers are one of your best sources for pricing information.

Let's say that you are a general contractor and that you build garages, additions, decks, and related home improvements. One way for you to establish the value of these home improvements is to consult with a licensed real estate appraiser. The appraiser can tell you what value the home improvement will have on an official appraisal of the property. This doesn't guarantee that the values given by the appraiser are the best prices for you to use, but they are an excellent reference.

Assume the appraiser gives you a value of $1,000 for a 10-×-10 deck. You could ask the appraiser to provide a written statement of value. Of course the value will be generic and may not apply to all conditions. Many factors affect the value of real estate and improvements, but the $1,000 figure will be a solid average. You can use this written statement as a sales tool when you sit down with customers. If you are willing and able to sell the deck for $800, you can show the customer how you are giving them a 20-percent discount off the average retail value of the deck.

You can also use the appraiser's figures as a piece of your puzzle, rather than the last word. By combining the appraised value with numbers given in pricing guides and your own estimate, you can come up with solid working numbers.

Newspaper ads

Newspaper ads can sometimes give you a feel for what your competitors are doing, but be careful. Advertising can be deceiving. When you look at an ad that offers to build a new garage for only $3,500 there is usually a catch or an angle. Maybe the concrete floor and footings are not included in the price. Maybe the quality of the construction is poor. Maybe your competitor has no idea how much money will be lost by selling a garage so cheaply. Watching other people's ads is a good idea, but you will have to investigate the advertiser to understand all the angles of the offer.

Combine methods

It is best to combine methods when you determine your pricing principles. Consult as much research material and as many resources as possible before you set your fees. Once you feel comfortable with your rates, test them. Ask customers to tell you their feelings. There is no feedback better than that of the people you serve.

PRESENT YOUR PRICES PROFESSIONALLY

Proper presentation is critical for business success. Even if your price is higher than the competition, you can still win the job with an effective presentation. There are many occasions when the low bidder does not get the job. As a contractor, you can set yourself apart from the crowd by using certain presentation methods. What are these methods? There are numerous ways to achieve an edge over the competition. Ways to win the bid battle include the way you dress, what you drive, your organizational skills, and much more.

When you learn to show the customer why you are worth more money, you are more likely to make the most money for your time and effort. People are often willing to pay a higher price to get what they want. You have to convince the customer to want you and your business, not the competition. How can you sway the customers your way? Well, let's study some methods that have proven effective over the years.

Mail mistakes

You might be surprised how many contractors mail estimates to customers and wait to be called to do the job. Many of these contractors never hear from the customers again. Mailing bids to potential customers is usually not the best way to get the job. When consumers spread estimates out and go over them, it is difficult to see differences other than price. You want to influence customers with the extras you can bring to the table, so you need a better method of presentation for your quote.

If you must mail your proposal, make sure you prepare a professional package. Use printed forms and stationary, not regular paper with your company name rubber-stamped on it. Use a heavy-weight paper and professional colors. Type your estimates and avoid using obvious correction methods to hide your typing errors. If you are mailing your price to the customer, you must make your mailing neat, well organized, attractive, professional, and convincing.

Phone facts

One of the worst ways to present a formal proposal is by phone. Telephones are great tools for prospecting and following up on estimates, but they are a poor vehicle with which to deliver initial estimates. When you call in your price, people can't see what you are giving them. They can't linger over the estimate and evaluate it like they can a written one. Chances are good that the customer may write down your price and then lose the piece of paper. Phone estimates also tend to make contractors look lazy, since they don't take the time to present a professional, written proposal. In general, use the phone to get leads, set appointments, and follow up on estimates. Don't use it to give prices and proposals.

Dress code

Whatever you wear, wear it well. Be neat and clean. Dress in a manner that makes you comfortable. If you are miserable in a three-piece suit, you will not project as well to the customer. How you dress will depend somewhat on your business. If you normally wear uniforms, wear your uniform when you present your proposal. Jeans are acceptable and so are boots, but both must be clean and neat. Avoid wearing tattered and stained clothing. You don't want the customer to be afraid to ask you to sit on the furniture.

When you are deciding on what to wear, consider who is your customer. If the customer is likely to be dressed casually, then dress casually. If you suspect a suit will fit in with your customer's attire, consider wearing a suit. Obviously, if you are going to have to crawl around under the house or in the attic, a suit would not be considered proper dress. Choose clothes that will blend in with the customer. If you dress too well, you might intimidate the customer. If your clothes and jewelry are too expensive, the customer will probably think you make too much money.

Your vehicle

What you drive says a lot about you. If you pull up in front of the average house in a high-priced sports car, you are sending signals to the customer. When the customer looks out the window and sees their plumber getting out of a fancy car, they are going to believe the rumors about the outrageous prices plumbers charge. If the same plumber crawls out of a beat-up, ancient truck, they may assume the plumber is not very successful.

Choosing the best vehicle for your sales calls is a lot like choosing proper clothing. You want a vehicle that will make the right statement. A clean van or pick-up truck should be fine for most contracting businesses. Cars are okay for sales calls, but stay away from the fancy ones unless you are dealing with a clientele that expects you to spend $40,000 for your work car.

Confidence

You must be confident in yourself, and you must create confidence in the mind of the consumer. If you can get the customer's confidence, you can almost always get the sale. You will gain your own confidence through experience, but you must learn how to build confidence with your customers.

Talking will often gain your customer's confidence. If you are able to sit down with customers and talk for an hour, your odds of getting the job increase greatly. By showing customers examples of your work you can build confidence. Letters of reference from past customers will help you establish trust. But, if you have the right personality and sales skills, you can create confidence by simply talking.

As a business owner you are also a sales professional, or at least you had better be. Unless you hire outside sales staff, you are the one who works with the customers. If you learn basic sales skills, you will have more work than the average contractor, even if your prices are higher.

KNOW YOUR COMPETITION

You must know your competition. How you price your services and materials will be affected by your competition. Your prices should be in the same ballpark as your reputable competitors. If your prices are too high, you won't get much work. If your prices are too low, you might be flooded with work and low profits, not to mention angry competitors.

To peg prices on complicated jobs, like room additions, you will probably have to learn a little at a time. When you are giving an estimate for this type of job you might be able to pry information on the prices of your competitiors out of the customer. In many cases, the customer will let you see the written quotes they have received, especially if you have good sales skills. By telling the customer you would like to review the other estimates for accuracy and to see if you can do any better on your price, they will generally let you see the quotes. This type of in-field research is invaluable. Once you can see your competitors' formal proposals it will be easy to refine your pricing structure.

USE EFFECTIVE ESTIMATING TECHNIQUES

By learning from your mistakes and successes, you can mold your own profitable estimating methods. Once you have ways that work, use them over and over again. You may have to alter your techniques to match them with specific customers, but the same basic principles that work for you in one sale will work in another.

How do you develop effective estimating techniques? You can learn the technical aspects of estimating by reading. Your estimating skills can be further perfected by referring to estimating handbooks and pricing guides. Much of what you learn will come from experience. Learning from your mistakes can be expensive, but you won't soon forget such costly lessons.

Organization is one of the most important factors in effective estimating. When you are organized you are more likely to complete a thorough estimate. Let's take a quick look at some ways you can improve your estimating style.

Keep notes

If you are sitting in your office and can't remember if the house had a 3-inch drain or a 4-inch drain, you are in trouble. The same would apply if you can't remember the size of the load-bearing beam. When you estimate, it is necessary for you to address and evaluate every aspect of a job. When you can't remember details, your estimate is unlikely to be a good one. Your price will either be too high to compensate for what you can't remember, or too low because you didn't have the facts you needed to make an accurate estimate. By keeping notes on all your potential jobs, your estimates will be much more accurate.

Look closely

You must look closely for items that will affect your estimated price. If you fail to notice that the electrical panel is not large enough to handle the load from the new work, you may have to absorb the cost of installing a new box. As a professional, it is your responsibility to know what needs to be done to complete the job. Not all customers are going to be willing to pay for extras that come up after the job is started. Keeping notes is important, but looking closely at existing conditions, plans, and specifications is essential to a good estimate.

Listen to the customer

Listen to the customer when you estimate a job. Not only will the customer appreciate your interest, you may gain valuable information. If the customer talks about roughing in plumbing for a future installation, make a note of it. Later you can solicit the customer for the additional work.

Keep copies

Keep copies of all your estimates. Some contractors fail to keep copies and are extremely embarrassed when they can't remember what they proposed to their customers. Keep copies of your notes and formal proposals. You never know when they will be needed.

Make files

Make files for each of your estimates. If you pile the estimates on your desk, not only will you have a cluttered desk, you may lose the estimate information. If a would-be customer calls to go over an estimate, you don't want to have to scramble through a stack of papers to locate the right estimate. It is much easier to open the file drawer and produce the file quickly and professionally.

Sell from the start

Good estimators begin selling right from the start. From the moment these estimators make contact with the customer, they are selling. Many good salespeople can talk the customer into adding additional features to their proposed plans. This technique can serve you well in two ways. First, you get more work. Second, if you can get the customer to change the description of work, you eliminate the validity of any previous estimates. This gives you a better shot at getting the job.

THE FEAR FACTOR CAN SELL JOBS

If you prepare yourself with stories to tell, you can increase your in-home sales. People are often nervous about the dependability, ability, and performance of contractors. This nervousness stems from fear created by the media. When people read a news story about how a homeowner was cheated out of money by an unscrupulous contractor, they become concerned. This type of news can hurt business for contractors, but it can also be your ace in the hole.

Once you have developed a portfolio of contracting-related horror stories, you can use the fear factor to sell more jobs. People generally assume you wouldn't be educating them in the risks that are present with contractors if you were one of the bad guys. Automatically, when you begin warning the consumer, you are building the image of one of the good guys.

After telling your stories, tell the customer how you operate to put their mind at ease. By showing the consumer what could happen, and why it won't if they work with you, you're on your way to signing a new deal.

13 Schedules, budgets, and job costing

Job costing, budgets, and schedules all play a vital role in the development of a contracting business. Without these elements, your business will be running in the dark. You can't afford to blindly wander through the business arena. If you do, your competitors will knock you out of the game.

Schedules should be used for many facets of your business. You should have a production schedule, a daily schedule, and a delivery schedule, to name a few.

Budgets should also be used for multiple aspects of your business. You should have an overall business budget, an advertising budget, and an inventory budget. Additionally, you will need budgets for individual jobs and the projected growth of your business.

Without the skills to make and follow schedules and budgets, a business owner is likely to make mistakes and lose money. One way to tell if a company is losing money is job costing. As jobs are in progress and as they are completed, job costing should be done and monitored. This practice allows the business owner to determine future pricing, forecast cash-flow needs, and derive an overall view of the company's on-the-job performance.

ESTABLISH PRODUCTION SCHEDULES

Production schedules will make your business more organized and profitable (FIG. 13-1). Production schedules allow you to plan and track your workload. Without production schedules, you will have trouble completing your jobs on time. When your jobs run past their completion dates, you will have angry customers.

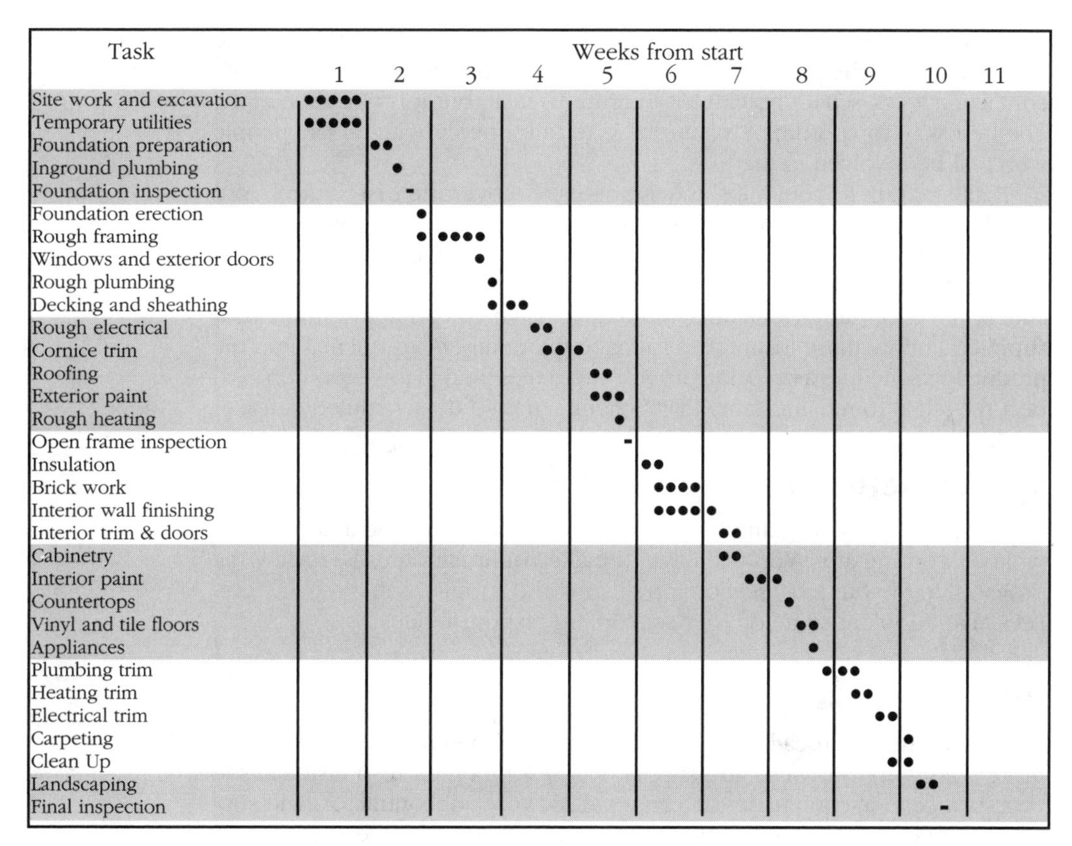

13-1 Typical construction work schedule. Department of Agriculture

Rough up your production schedules on a computer or on notepaper. Start with just one job. Once you have a schedule for the first job, scheduling the remainder of your jobs will be easier. If your jobs overlap, you may have to schedule multiple jobs at the same time.

Start by putting the job name and address on the schedule. Lay out the schedule with headings for each phase of work. Create spaces where you can fill in dates for the various work phases. You will need spaces for many dates. The first space should contain your anticipated start date. Another space should contain the date work is actually started. You should have an entry for the estimated time needed to complete the task. Then you should have a spot to enter the actual completion date. In addition to your start and finish dates, allow space to write in dates for material deliveries.

The production schedule should have provisions for listing the names and phone numbers of the subcontractors and suppliers on which you will depend. Write these names and numbers next to each work phase and delivery. Having the names and numbers on the schedule will make it easier to make follow-up confirmation calls.

Now all you have to do is fill in the appropriate dates. Choosing the dates for your schedule may not be simple. Attempting to work out non-conflicting work schedules can be an arduous task, but it is not impossible. The best way to coordinate your jobs is to talk openly with all the people who will be involved in the jobs.

Call and talk to your subs and suppliers. Go over the first draft of your schedule with them. Ask if they will be available on the dates you want to have them on the job. If you don't know how long it will take the plumber to rough-in the job, ask. When you are concerned about how long it will take to have roof trusses manufactured and delivered, get advice from your supplier. Follow these techniques to make the numbers you pencil into the production schedule as accurate as you can make them. However, don't expect the dates to remain static. Plans change and so do scheduled dates.

STAY ON SCHEDULE

Staying on schedule is going to take effort. You cannot maintain a schedule without working at it. You will have to make confirmation calls, review the progression of your jobs, stay on top of subs and suppliers, and much more. Let's take a look at some of your scheduling responsibilities.

Confirmation calls

Making confirmation calls is essential to staying on schedule. If you don't make follow-up calls, subs and suppliers may forget their responsibilities to you. As a general contractor you cannot allow your subcontractors and suppliers to overlook their obligations. It is your job to assemble and manage all the elements of a successful job.

Calling suppliers to confirm orders and deliveries can be done during normal business hours, but this isn't always true with subcontractors. Many subcontractors work in the field. These people don't spend their days sitting next to a phone. To get through to busy subs, you may have to spend time at night making your calls. This can get old, fast.

The nature of your business will determine how long you have to work. General contractors often work from dawn into the late evening. You can reduce the need to work so many hours if you have your subcontractors check in with you at specific times. By asking your subs to call you when they take breaks or lunch, you can eliminate some of the night calls.

Even with the best of efforts, making follow-up calls can become a pain. It may be tempting to let some of your calls slide, but don't do it. If you don't confirm your schedule, the schedule will be useless.

Track production

Tracking production is one way to keep your schedule accurate. Schedules require adjustments. To make these adjustments you must keep track of your production. If the drywall contractor is running behind, you will have to reschedule the painter. If a special-order item, like a custom window, is past-due, you will have to adjust the schedule. To make these adjustments you must keep your finger on the pulse of the job. This generally means checking each job's status on a daily basis.

After you check your jobs, adjust the schedule to allow for the field production.

Keep subcontractors and suppliers in line

Keeping subcontractors and suppliers in line is another responsibility associated with staying on schedule. If subs or suppliers don't fulfill their obligations in a timely fashion, your schedule will be blown out of the water. You must control your subs and suppliers to stay on schedule.

If you follow my earlier advice and include start and finish dates in your subcontractor agreements, you are on the right track. Including a daily penalty charge for every day the job is not finished will give you an even better chance of staying on schedule (FIG. 13-2). Couple these two contract clauses with persistent prodding, when needed, and you should do fine. If all else fails, exercise your ejection clause and bring in a new sub or supplier to fulfill your needs.

Juggling

There will be times when it will be necessary to juggle your schedule. It is one thing to juggle to accommodate unforeseen production changes, but it is quite another thing to juggle for cash flow or greed. Before we go on, let's establish the difference between the two types of juggling.

It is necessary and acceptable to juggle your jobs when materials are late or subcontractors don't perform as expected. This type of juggling usually extends the estimated completion date, but it might mean moving up the completion date. For example, if you thought it was going to take 10 days to complete the drywall work, and it only took 7 days, you can move your schedule up by 3 days. This, of course, is your choice. You also could let the existing schedule remain intact.

Juggling your work schedule to generate cash flow or to take on more work is dangerous. If you move your crews from job to job, in order to start more jobs or to collect more cash draws, you will probably be out of business within a year. This type of job juggling will give your business a bad reputation and will result in more trouble than it is worth.

ADJUST FOR UNFORESEEN SCHEDULING OBSTACLES

After being in the business for awhile, you start to anticipate the unexpected. Seasoned contractors are so good at projecting problems that they rarely have them. Unfortunately contractors that are new to the business suffer through some tough times and pay a price for their lack of experience.

Many people assume that because they know their trade, they know their business. Knowing how to do your job as a tradesperson is not the same as knowing how to run a business. You might be very competent at estimating the time you will need for the work you will do yourself. If you're are not an experienced contractor, however, you may not know what allowances to make for subcontractors and suppliers. These two variables can destroy your scheduling plans.

During my many years in business I have never ceased to be amazed at how some people conduct business. I am a master plumber, licensed remodeler, licensed builder, and licensed real estate broker. I have been involved with property management, remodeling, plumbing, and building. During these experiences I have seen a number of different trades function. I have also gotten a good feel for how long it should take to rough-in the electrical wiring for a home or to install a new roof.

After building up to 60 houses a year, I feel I am well qualified to project the time needed for various trades to do their jobs. You would think I could estimate these work requirements with extreme accuracy, but it doesn't always work out that way.

I know that I can rough-in the plumbing for an average 2½-bath house in two days, if I have a helper. It will take about half a day to test the rough-in and have it inspected. Then I will need one day and a helper to set all of the plumbing fixtures on the trim-out. Now this is what I can and have done. So shouldn't most experienced plumbers be able to perform the work in a similar amount of time. It seems to me they should, but it is surprising how long some plumbers can take to do the same job.

I have hired employees and subcontractors to perform plumbing work. I have had some very good plumbers. During my prime, I rarely found anyone who was substantially faster than I was, but I constantly found plumbers who were extremely slow. Their slowness is caused by laziness, inexperience, lack of focus, and a number of other reasons.

For the same work I can accomplish, with a helper, in two and a half days, some crews take a full five days. As a contractor, you cannot assume that all subcontractors will complete their tasks in the same period of time. It is very possible that one sub will need twice as long to get the job done as another. As a contractor, you have to find the secret to meandering through the maze of subcontractors and suppliers.

As you gain experience and get to know your subcontractors and suppliers, you will become more adept at adjusting your schedule. You can almost become a mind reader when it comes to how your schedule will need to change. This is not to say that you can avoid all of your scheduling problems, but you can learn to take control of them.

PROJECT YOUR JOB BUDGET

Job budgets are the basis you use to set your profit goals. If you miscalculate your job budgets, you will not make the profits you expect. Occasionally, you may make more money than you expected. In most cases, you will leave the job with less money than you projected.

For some people, the term budget is a dirty word. This mind-set extends from personal budgets right into business budgets. While it may not

Subcontractor Schedule

Type of service	*Vendor name*	*Phone number*	*Date scheduled*
Site work			
Footings			
Concrete			
Foundation			
Waterproofing			
Masonry			
Framing			
Roofing			
Siding			
Exterior trim			
Gutters			
Pest control			
Plumbing/R-I			
HVAC/R-I			
Electrical/R-I			
Insulation			
Drywall			
Painter			
Wallpaper			
Tile			
Cabinets			
Countertops			
Interior trim			
Floor covering			
Plumbing/final			
HVAC/final			
Electrical/final			
Cleaning			
Paving			
Landscaping			

Notes/Changes

____

____

13-2 Subcontractor schedule form.

be fun to follow the guidelines of a financial budget, budgets are often the only way to reach your monetary goals.

If you don't project a budget for each of your jobs, you are sure to be disappointed with some of your income results. Budgets may not be enough to keep you from losing money on a job, but they can help you spot your problems and avoid making the same mistake twice.

If you sell a job too cheaply, setting a budget is not going to solve your problem. The best budget going will not compensate for a poorly-estimated job. However, by budgeting your jobs you can evaluate your results and price future jobs to keep your profit margin where you want it.

To make a job budget you must project and list all anticipated job costs. For some contractors, thinking of all the expenses that will be incurred in a job is a chore they cannot fathom. These people don't have the natural ability to put all the steps of a job into chronological order. This doesn't mean the contractors can't learn to forecast a budget, they just are not naturally talented in this aspect of running a business. Building a job budget is not difficult if you proceed in a logical way.

Just like you build a production schedule, label your form with the job name and address. Continue by listing all aspects of the work phases. For example, if you are going to forecast your plumbing expenses, include spaces for sewer work, water-service work, groundwork, rough-in work, and trim work.

Most contractors don't have problems listing the large phases of work, like carpentry, plumbing, heating, electrical, painting, and so on. However, many business owners neglect to list the smaller phases of work required on a job. These minor phases can add up to quite a lot of money.

Consider all of your anticipated costs (FIG. 13-3). Let's look at some of the costs that might be incurred in a major remodeling project that often go undetected in a budget. These costs could include:

- Demolition costs
- Trash-disposal fees
- Clean-up costs
- Drafting or architectural fees
- The cost of code-enforcement permits
- Administration costs
- Office overhead expenses
- Charges for electrical power
- Inspection fees
- Field supervisors
- Advertising costs incurred to obtain the job
- Time spent estimating and selling the job
- Potential warranty work

This list could go on for pages. There are many costs associated with doing a job that will not show up on most contractors' budgets. If these soft costs are left out, the profits from the job will shrink. You must be thorough when you itemize your job expenses. If you are not, the budget is bogus.

Cost Projections

Item/Phase	*Labor*	*Material*	*Total*
Plans			
Specifications			
Permits			
Trash container deposit			
Trash container delivery			
Demolition			
Dump fees			
Rough plumbing			
Rough electrical			
Rough heating/ac			
Subfloor			
Insulation			
Drywall			
Ceramic tile			
Linen closet			
Baseboard trim			
Window trim			
Door trim			
Paint/wallpaper			
Underlayment			
Finish floor covering			
Linen closet shelves			
Closet door & hardware			
Main door hardware			
Wall cabinets			
Base cabinets			
Countertops			
Plumbing fixtures			
Trim plumbing material			
Final plumbing			
Shower enclosure			
Subtotal			

13-3a Cost projection form (page 1).

Item/Phase	*Labor*	*Material*	*Total*
Light fixtures			
Trim electrical material			
Final electrical			
Trim heating/ac material			
Final heating/ac			
Bathroom accessories			
Clean up			
Trash container removal			
Window treatments			
Personal touches			
Financing expenses			
Miscellaneous expenses			
Unexpected expenses			
Margin of error			
Subtotal from first page			
Total estimated expense			

13-3b Cost projection form (page 2).

PREPARE ACCURATE TAKE-OFFS

To make sound bids, you must be adept at preparing accurate take-offs. It doesn't matter if you use a computerized estimating program or a pen and paper. You must get your facts straight. If you miss items on the take-off and get the job, you will lose money. If you overestimate the take-off, your price will be too high. An accurate take-off is instrumental to the success of winning a job.

What is a take-off?

A take-off is a list of items needed to do a job (TABLE 13-1). Take-offs are the result of reading blueprints or visiting the job site and making a list of everything you will need to do the job. Some estimators are wizards with take-offs. Others have a hard time trying to project all of their needs. If you can't learn how to do an accurate take-off, your venture into contracting is going to be rough.

Use take-off forms

You can reduce your risk of errors by using a take-off form. If you use a computerized estimating program, the computer files probably already contain forms (FIG. 13-4). You may want to customize the standard computer

Table 13-1 Materials Take-off List.

Item name or use of piece	*No. of pieces*	*Unit*	*Length in place*	*Size*	*Length*	*No. per length*	*Quantity*
1. Footers	45	Pc	1'5"	2×6	10'	7	7
2. Spreaders	30	Pc	1'4"	2×6	8'	6	5
3. Foundation post	15	Pc	3'0"	6×6	12'	4	4
4. Scabs	20	Pc	1'0"	1×6	8'	8	3
5. Girders	36	Pc	10'0"	2×6	10'	1	36
6. Joists	46	Pc	10'0"	2×6	10'	1	46
7. Joist splices	21	Pc	2'0"	1×6	8'	4	6
8. Block bridging	40	Pc	1'10⅜"	2×6	8'	4	10
9. Closers	12	Pc	10'0"	1×8	10'	1	12
10. Flooring	800	BF	RL	1×6	RL	—	—

Department of the Army

Job Take-Off Form

Job Name: ____________________

Job Address: ____________________

Item	**Quantity**	**Description**
2" pipe	100'	PVC
4" pipe	40'	PVC
4" clean-out w/plug	1	PVC
2" quarter-bend	4	PVC
2" coupling	3	PVC
4" eighth-bend	2	PVC
Glue	1 quart	PVC
Cleaner	1 quart	PVC
Primer	1 quart	PVC

13-4 Take-off form.

forms. Whether you are using stock computer forms or making your own forms, you must be sure they are comprehensive.

Take-off forms should list every item you might use for various types of jobs. In addition, there should be blank spaces on the form that will allow you to fill in specialty items.

As an example, assume you are a plumber. You would have several types of take-off forms. You might have one for doing the rough-in plumbing on a new house. You could also have one for the finish work in the house. It's conceivable that you would have a different form to list materials for any underground plumbing and the water service and sewer. Of course, you could have all phases of your work on one form, but that could get confusing.

When the time comes to figure your material needs for installing a sewer, you can go right down your form and fill in the quantities of each item you will need. For instance, you might need 70 feet of 4-inch pipe and 3 clean-out assemblies. You might also need some wyes, eighth bends, and caps. If you are working with plastic pipe, you will need glue and probably a cleaner. This list could go on, but you get the idea.

The advantage of using take-off forms is that you are prompted on items you might otherwise forget. However, don't get into such a routine that you only look for items on your form. A job might require something that you haven't yet put on the form. Forms help, but there is no substitute for thoroughness.

Videotape the job

Some estimators use video recorders to videotape the job. While this fad has not taken a strong hold on the contracting industry, it does seem to be on the rise. This is especially true with remodeling and rehab jobs. By videotaping the existing conditions, you can refer back to the tape when you complete your take-off form. If your field notes aren't clear on the type of existing floor covering, you can watch the tape. While video recorders aren't for everyone, they may work well for you.

Keep track of what you count

If you are doing a take-off form on a large set of plans, say a shopping mall, it can be tedious work. The last thing you need to have happen is to lose your place or forget what you've already counted. To avoid this problem, mark each item on the plans as you count it.

PROJECT YOUR BUSINESS BUDGET

Projecting your business budget is similar to forecasting a job budget. In both instances you must account for all anticipated expenses. Unlike a job budget, a business budget will have to include projections for a much longer period of time. This extended time can make the creation of a good business budget more difficult to master.

Developing a business budget that will work requires extensive thought. You must consider your immediate needs and project your future needs. Your desires will come into play with a business budget. For example, if you want to retire in 20 years, you must make an allowance for this desire in your budget. Let's take a look at some of the factors you should incorporate into your business budget.

Salary

Many business owners don't pay themselves a set salary. They make as much money as they can and use the money they need. This works for some people, but it is not the way to make a business budget. When casting your business budget, establish a set amount for your salary. You may not always be able to take as much money for your salary as the budget reflects, but you need to plug in an income figure.

When you set your salary, don't forget your responsibility for income taxes. You have to look at your net-income needs and the gross-income requirements. There is a substantial difference between your net and gross needs.

Office expenses

Office expenses can cause your budget to balloon. Think about all the various expenses that you incur to keep your office running: rent, utilities, phone bills, cleaning, equipment rentals, office supplies, furniture, and much more. The cost of your office expenses can account for a high percentage of your annual expenses.

Many contractors have excessive office expenses. When you build your business budget, check over your office expenses. You will likely find ways to save money.

Vehicle expenses

Almost every contracting business is affected by vehicle expenses. The cost of trucks, cars, fuel, tires, and similar expenses can amount to thousands of dollars. Of course, your vehicle expenses will be related directly to the size and structure of your business. If you rely on subcontractors for all of your field work, your vehicle expenses will be minimal. However, if you have a fleet of trucks and an army of employees, vehicle expenses can be astronomical.

Tool and equipment expenses

There are two types of tool and equipment expenses. There is the cost of initial acquisitions and the cost of replacement. Even if you have all the tools and equipment you need, you have to budget for replacement costs. Your hardware may be broken, stolen, or worn out—sooner or later it will need to be replaced. If you are not prepared financially for these replacements, you can find yourself unable to continue doing business. Take a look at your tools and equipment. Estimate the life expectancy of your hardware and put a replacement figure in your budget.

Employee expenses

When you look at employee expenses, you have to look much deeper than the hourly rates earned by your employees. You must look at employer taxes and employee benefits. If you provide insurance benefits for your employees, you are spending serious money. Paid vacations are another major expense. Sick leave, paid holidays, and similar employee benefits can amount quickly to thousands of dollars.

If your business is top-heavy with employee expenses, you might do well to consider engaging independent contractors.

Insurance expenses

Insurance needs vary depending on the nature of various businesses. Liability insurance is a need for all businesses. With employees, worker's compensation insurance can be an expensive factor in the business budget. Insurance to protect against theft and fire is another common business need. When a company is properly insured, the expense of the insurance can account for much of the total business budget.

Advertising expenses

Advertising expenses are one of the few expenses from which businesses see a direct gain. When used efficiently, advertising will pay for itself and then some. However, a business owner can never be sure of the effectiveness of his advertising.

While you cannot be sure of your advertising results, you can project how much you are willing to spend to generate new business. Many companies dedicate a percentage of their gross sales to advertising. Most small companies pick a dollar amount rather than a percentage. Whichever method you use, make sure to include your anticipated advertising costs in your business budget.

Loan expenses

Don't overlook loan expenses when you build your budget. This is especially true if you will use short-term, interest-only loans. The cost of these loans may not register as an expense since the interest is usually paid in a lump-sum and not on a monthly basis. Don't forget to include the fees you will incur with your business financing.

Taxes

Taxes are another expense that generally is left out of a business budget, but shouldn't be. The impact of taxes can cause severe stress on a business.

Unless regular tax deposits are made, the burden of coming up with enough money to satisfy the tax authorities can be overwhelming.

To avoid being caught in a cash bind, include your estimated tax liabilities on your individual job budget projections. If you are unable to forecast your taxes, consult a tax professional. You cannot afford to be left high and dry on tax day.

Growth expenses

To see business growth, you must plan for growth expenses. Few businesses grow without a plan. Consider your intent for company growth when you build your budget, and include provisions for expansion capital.

Retirement goals

Retirement goals frequently get pushed to the back of the list. When money is tight, it is hard to invest in retirement plans. The impulse is to fight the fires of today and worry about retirement later. Unfortunately, age and retirement creep up on all of us, often sooner than we plan.

New business owners often see the business they are building as their retirement. In some cases their beliefs are founded, but more often the business is not enough for retirement. In fact, many a new business won't be in operation when the owner desires to retire. For these reasons, investments in your business are not enough to guarantee your safe retirement.

To build a suitable retirement portfolio you need a plan. The plan will normally involve diversified investments: stocks, bonds, real estate, coins, art, and a number of other possibilities. It is a wise idea for most people to talk with professionals in order to design a viable retirement plan. Once you have laid a solid path to retirement, include the funds for your retirement in your business budget.

As you build your business budget, you may discover other items that should be included. Remember that your budget is your business map.

MAINTAIN YOUR BUDGET REQUIREMENTS

Having a budget will do you little good unless you have the discipline and skills to maintaining it. It helps to have a natural ability to maintain budget constraints, but if you don't, you can learn the skills and discipline.

Self discipline is a factor that influences all aspects of our lives. It is the quality that allows us to live within the laws of society. Some people have strong discipline for certain aspects of life and no discipline in others. Compulsive gamblers may discipline themselves to go to work on time and maintain an average life, but lose it when it comes to gambling. People with an obsession for food may be perfectly normal, until food is set in front of them. Most of us have some weak areas in our self-discipline. Your job as a business owner is to find your weak areas and reinforce them.

All of your potential weaknesses must be evaluated. If you are quick to give in to impulse buying, you must set rules for yourself. Make yourself wait a reasonable period of time before you buy that new tool or office equipment. Wait and see if you need the item or if you only want the item.

Examine all of your business habits. Are you going to lunch at expensive restaurants on a regular basis? If you are, you may be spending enough to pay for a more justified business expense. Document all of your spending for the next two weeks; include every item you spend a dime on. At the end of the two weeks, go over your list of expenditures. Compare these expenses with your budget. My guess is that you will find several areas where you are spending money that wasn't budgeted. This is a danger signal. You can't allow yourself to run rampant with your company's financial resources. If you do, your business will fail.

After you tag all the unbudgeted expenses, decide if you will continue to make similar purchases. If you come to the conclusion that your expenses are justified, include them in the budget. If the spending habits are not necessary, eliminate them. Periodically repeat the two-week test. By monitoring and refining your budget, you can stay within its framework.

ASSESS FUTURE JOB PRICING WITH JOB COSTING

To make the most of your business, you must obtain the best price possible for your services. Sometimes this will mean increasing your prices; sometimes it will mean lowering your fees. You can develop a strong sense of direction by reviewing and assessing your profit performance on past jobs. Job costing will give you the information you need to make sound decisions on your pricing structure.

Job costing is the act of adding up all the costs incurred to complete a job. Depending upon your methods, job costing may account only for labor and materials used on the job. In more sophisticated reports, the numbers will include the cost of overhead and operation expenses (FIG. 13-5).

Your routine business expenses must be factored into the cost of jobs, but you don't have to include it in individual job-cost reports. Instead, you can use a percentage of your on-job profit to defer your other costs of doing business. Let's see how each of these methods can be used.

Simple job costing

Simple job costing should include all expenses incurred for a specific job, excluding overhead expenses. Overhead expenses might include: office rent and utilities, general insurance, advertising, and so forth.

To do a simple job cost, you need nothing more than pencil, paper, and information. First, list all the materials you used on the job. Then, price these materials, based on what you paid for them. Next, calculate the cost of all the labor that went into the job. Then include soft costs: permit fees, the cost of blueprints, and so on.

Job Cost Log

Job name ______________________________

Mechanic ______________________________

Truck ______________________________

Date ______________________________

Item	*Quantity*	*Size*

13-5 Job cost log.

Once you have listed all the costs for the job, check over the list for omissions. When you are sure you haven't forgotten anything, add up the total of the costs. Subtract the total of your costs from the money you received for doing the job, and you've got your gross profit.

Let's say you sold the job for $10,000. The total of your on-job expenses was $8,000. This means your gross profit was $2,000 or 20 percent. Does this mean you made a profit of $2,000 on the job? No. The gross profit of $2,000 doesn't reflect your off-job expenses, which must also be subtracted from the gross profit.

Let's say that you had four jobs of equal value and equal effort running at the same time. For this example, assume that you started and finished all four jobs in one month. Further, let's say that your off-job expenses average $4,000 a month. In this simple example, you would divide your off-job expenses by four to obtain the percentage of these expenses that should be charged to each job. In this case, your off-job expenses equal $1,000, or 25 percent of your overhead was attributed to the job in question. After doing this math, you find that your net profit on the job was $1,000 or 10 percent.

In real life, it is not as simple to account for overhead expenses. It is not uncommon to have several jobs running at once and many jobs overlapping each other. This complicates the division of overhead between jobs. Some businesses project a percentage of their overhead and apply it to every job. Other businesses use complex methods to allocate the exact cost of their overhead to each job. How you do your job costing is up to you, but you must subtract the off-job expenses to get a true impression of your profit margin.

Percentage job costing

Percentage job costing is another common method for determining profits and losses. To perform this task, duplicate the instructions given for simple job costing. Once this is done, subtract a percentage of the contract price for the job. The percentage you subtract will represent an estimate of all your overhead expenses.

Using the same numbers as in the earlier example, let's see how this math would work. In the first step you determined a gross profit of $2,000, excluding overhead expenses. The job sold for $10,000. Let's say that after testing you estimate your overhead factor to be 10 percent. So, find 10 percent of the contract price and deduct it from the gross profit. In this case, 10 percent is $1,000 When you subtract $1,000 from $2,000, you arrive at a net profit of $1,000.

What percentage should you use? The percentage applied for overhead will depend on your business structure. Some businesses carry heavy overhead expenses and others are streamlined to maximum efficiency. You will have to experiment with your personal circumstances to determine what percentage will cover your off-job expenses.

Adjust your prices

Adjusting your prices will be much easier when you work with the results of accurate job-costing reports. If, after reviewing the job costs on three similar jobs, you see a pattern, you can determine if you are charging enough, too much, or not enough.

Let's look at an example of a company that specializes in building decks. The company builds all kinds of decks in various sizes. The business owners, Larry and Ann, have carefully plotted the percentage of overhead that must be applied to each job. While the company is staying busy enough, it could use some additional business. The owners want to find out if it is feasible to lower prices to increase volume. Let's see how the job-cost reports on various decks will prove helpful in this decision.

Larry and Ann have sold a lot of deck jobs. The most popular size for decks has been around 200 square feet. To decide on a pricing path, Larry and Ann pull job-cost reports from various jobs. The jobs under investigation range from 100 square feet to 400 square feet.

When they review the job-cost data, the business owners see that they make a 20-percent profit on decks with 100 square feet. The profit percentage rises to 30 percent for decks containing 200 square feet and to 35 percent on decks in the 400-square-feet category.

Larry and Ann have built their business around a projected profit of 20 percent on all jobs. Based on this, they cannot afford to lower the prices of their small decks. However, there is room to discount the prices on larger decks.

After seeing how their profit escalates with the size of the job, Larry and Ann decide on a marketing strategy. They will advertise their most sought-after decks, the 200-square-feet variety, with a 10-percent discount. This advertising should result in increased sales, while maintaining the desired 20-percent profit.

This is only one way of using job-cost reports to adjust your pricing. You might find that you need to raise your prices. The key is being able to determine how your business is doing in the area of net profits. Once you have concrete facts, you can adjust your prices to satisfy your goals.

TRACK YOUR PROFITABILITY

We have just seen one example of how business owners can assess their job profits by utilizing job-cost reports. In the earlier example, Larry and Ann were looking for ways to adjust their pricing. They weren't looking specifically to track their profits. However, since they had been tracking their profits on past jobs, they were able to use the information to make an informed decision on their pricing.

Job costing will allow you to monitor all of your jobs for maximum profit potential. It is easy to say you want to make a gross profit of 20 percent on every job, but it is not so easy to accomplish. Some jobs yield a higher profit potential than others. This is where job costing can lead you in the right direction.

To see how different jobs produce various results, let's consider the circumstances of a general contractor. This contractor deals only in residential jobs. The jobs range from minor remodeling to major renovations. New construction, such as garages and additions, also make up a part of his business.

The general contractor has allocated $15,000 for advertising over the next year. Knowing how much he has to spend on advertising, the contractor must decide how to spend it.

By job costing, the general contractor learns how much he made or lost. It allows him to outline a plan for his business. Advertising is more effective when it is the result of a careful study of job-cost reports. Budget distributions are easier to establish when you have detailed job-cost reports to study. There is almost no limit to how accurate job-cost reports can help your business.

ACCURATE JOB COSTING IS ESSENTIAL TO LONG-TERM SUCCESS

To endure the test of time, businesses must be financially sound and flexible. Job costing will show you where you stand financially. You will discover categories that were left out of your budget. Job-cost reports can bring the reality of overhead expenses to light. While you may never have considered charging a portion of your rent to a job, you will see that you must.

The information you gain from accurate job-cost reports will make you a better estimator. If you lost money or made a minimal profit on your last job, the next job should go better. Since you will be able to pinpoint your problems with job costing, you can adjust your next estimate to protect yourself.

Build in a margin of error

If you think you are going to need 100 sheets of plywood, add a little to your count. How much you add will depend on the size of the job. A lot of estimators build in a float figure of between 3 and 5 percent. Some contractors add 10 percent to their figures. Unless the job is small, I think a 10-percent add-on might cause you to lose the bid.

Of course, much of your cushion for mistakes will depend on your ability to accurately complete a take-off form. If you are good with take-offs, a small percentage for oversights will be sufficient. If you always seem to get on the job and run short of materials, you will need a higher slush pile.

Keep records

Don't throw away your take-off forms. When the job is done, compare the material actually used with what you estimated on your take-off form. This will help you see where your money is going, and make you a better estimator. Tracking your jobs and comparing final counts with original estimates will help you refine your bidding techniques and win more jobs.

14
Customer relations

Most contractors recognize the need for customers, but many underestimate the importance of public relations. By being ignorant of, or blind to, the importance of public relations, many contractors lose business. The business they lose goes past without leaving tracks. These unfortunate contractors don't know why they lose business, only that their business volume is down.

If you don't want to fall into the category of contractors dazed by a loss of business, hone your public relation skills. When you improve these skills, you will see an improvement in your business. The improvement may not be obvious, but if you look closely, you will see it.

MEET YOUR CUSTOMERS ON THEIR LEVEL

When you are working with people, you want those people to be as comfortable as possible. For this reason, you have to be able to change clothes and personalities. The most successful salespeople are the ones who are chameleons. These gifted people can move up and down the social ladder to serve any prospect that comes along. This flexibility provides an advantage the average salesperson doesn't possess. Being able to assume the personality of a customer makes it easier to close a deal.

Fitting in, with what you wear and drive, is another key to sales success. When you are dealing with customers, attempt to fit in with the customers. This might mean wearing a suit in the morning and jeans in the afternoon. You might find it advantageous to switch from your family car to your pickup truck. The more effort you put into blending in with the customer, the more sales you will make.

Customers are what give your business value. How you treat existing customers can influence the long-range success of your business. If you alienate customers, they won't give you return business or referrals. It is much less expensive to keep good customers than it is to find them. Once you have established a customer base, work hard to maintain it.

QUALIFY YOUR CUSTOMERS

Don't be afraid to ask your customer questions. While it is true that the customer is hiring you, you have the right to know a little about your customer. One of the first questions you may be concerned with is the customer's ability to pay for the work being requested. This may seem like silly curiosity, but it is not.

It is not uncommon to have customers who do not pay their bills. The reasons for nonpayment are extensive, but the end result is that you do not get paid. No business owner can afford to work for free. Let's look at some of the reasons you may not get paid by your customers.

Tenants

When there is a problem in a rented home, tenants will sometimes call you and request that the problem be rectified. Some of these calls will be of an urgent nature, others mundane inconveniences. Since the tenants don't own the property, they may expect you to bill the property owner. Since the landlord didn't request the service, it can be hard to collect from him. The landlord may feel that since the tenants acted on their own, they should be responsible for the bill. Either way, it will be difficult to collect your fees.

In a situation like this, at the best you will waste time trying to get someone to pay the bill. At the worst, you won't get paid. Most service repairs don't involve large sums of money. Going to court to collect for a 1-hour service call is a waste of energy and money. Tenants and landlords know you are not likely to take legal action to collect the money due you. This assumption on their part makes it even harder for you to keep your accounts receivable current.

You can stay out of this tangled web by qualifying the customer. When a customer calls for service, ask if they own the property. If they don't, inform the caller that you will not respond without authorization from the property owner. Once you are dealing with the paying party, have them sign a work order before work commences. The work order should detail all aspects of the job, including how and when you will be paid. This written agreement will carry more weight than a handshake.

Loan denial

Repair companies are not the only ones at risk. If you specialize in large jobs, like building additions, your customers will usually be dependent on borrowed funds to pay you. If the loan isn't approved, there will not be enough money to settle your bill. Contractors, especially new contractors, are often anxious to work. When a contract is signed for a big job, some contractors will start the job immediately, before the loan has been arranged. This is bad business.

To avoid getting stuck on a big job, ask the customer to show evidence of the money required to do the job. This could amount to seeing a loan agreement or a bank statement for the customer's account. You may feel awkward asking to see proof of available funds, but you will feel worse if you don't get paid.

Deadbeats

Avoiding deadbeats can be difficult; they are not always easy to identify. To protect yourself from this undesirable group of customers, get permission to run a credit check on your customers. This is good business. You never know when the sweetest, most trusting person is going to turn out to be a bad debt.

Your fault

Sometimes nonpayment will be your fault. If you have not made the customer happy, you might have trouble collecting your cash. When you qualify your customers, try to read them for trouble signs. If you feel friction at the meeting, perhaps you should pass on the job and look for another customer.

Death

Death is always a good excuse for not paying your bills. Of course, death is no laughing matter, but neither is not getting paid. While you can't avoid a customer's demise with qualifying, you can make arrangements in your contract to cover the death contingency. Ask your attorney for help drafting a clause that will hold the heirs and estate responsible for your fees if the client passes away.

Bankruptcy

If a customer owes you money and files for bankruptcy protection, your chances of being paid are all but nonexistent. During the qualifying stage you can screen the customer's credit rating and financial strength. If the customer is financially healthy when you start the job, there is limited risk of losing your money in the bankruptcy courts.

Insurance

Some customers believe the work they are requesting will be covered by their insurance policy. This situation could arise from a flood, a tree falling on a house, or whatever. Insurance companies don't always agree to pay for damages. If your customer tells you that you will be paid by an insurance company, talk to the insurance company before you do any work.

There are other reasons for qualifying your customers. The reasons are as diverse as the companies doing business. If asking questions doesn't intimidate you, you can learn a lot about your customers. The more you know in the beginning, the better the job will go.

SIMPLE SALES TACTICS SELL MORE JOBS

Professional salespeople spend time getting to know their prospects. Call it breaking the ice, probing, or whatever you want, but you will be more successful if you show an interest in your prospects.

Gain the customer's confidence and the job is yours

To get the confidence of your customer, you have to spend time talking. Talk about the customer's children, grandchildren, hobbies, car, or anything else that seems appropriate. You can get a good idea of how to start the conversation by looking at the pictures in the customer's home. If you see pictures of horses, talk about horses. If you see pictures of flowers, talk about flowers. Always allow the customer to lead the conversation. You can't afford to take a chance on saying the wrong thing.

After you have talked with the customer for awhile, talk a little business. If you seem to be losing the customer, go back to talking about a friendly subject. Tell stories about your life. Let the customer get to know you. Once you have settled in with the customer, getting the sale will be much easier.

Arrange face-to-face meetings

Whenever possible, arrange a meeting to go over your estimate with potential customers. If you mail or call in the price for a job, you have little opportunity to pitch the advantages of your company. A sit-down meeting will always produce more sales than an estimate that is called in or mailed.

Follow up on estimates

Following up on your estimate is also important. Most contractors don't check the status of their estimates. If you do a follow-up call, you have a second chance to sell the customer. Even if you don't get the job, you can get feedback on why you didn't get it. If you find out what you are doing wrong, you can amend your tactics and build a better business.

Before-and-after photos sell jobs

When you have the opportunity, take photos of your jobs before, during, and after completion. These photos build credibility and can be used as a part of your reference package. Photographs are also good for giving new customers ideas.

Create an album of all your before-and-after shots. Take the album with you on all of your sales calls and don't hesitate to show it to potential customers. You can use the photos to point out methods you use that your competition doesn't. For example, if you use pressure-treated lumber for sill plates, take a color photograph of the lumber. Show the photograph to your sales prospect and point out the advantages of using the pressure-treated lumber. Photos will help you win more jobs.

PRESENT YOUR PROPOSAL PROFESSIONALLY

Presenting your proposal is a critical part of making a sale. Like I said earlier, a sit-down meeting is a much more effective way of presenting your proposal than a phone call. But, there is more to presenting your proposal than handing it to a prospect. If you handle your sales presentation well, you can leave with a contract in your hand. I have sold additions for more than $20,000 on the spot, in less than two hours.

To present your proposal, arrange for all the buying power to be present. This normally means having husband and wife both at the table. You don't want to have your closing of the sale blocked by the old line, "I'll talk it over with my spouse and get back to you." Set the meeting for a time and place where the customer will be at ease. The customer's home is a good place to meet. Pick a time that will be conducive to an uninterrupted meeting. Don't set the meeting close to dinner time. Allow up to two hours to close large sales. I've found that if I can stay in front of the buying power for between one and half hours and two hours, I'll get the sale.

When you present your proposal, pitch your good points. Tell the prospect how your work is better than average. Don't name names and criticize specific competitors. This is a cheap shot and will not be respected by most prospects. You should study sales techniques and practice your presentation skills. When you become competent in sales methods, your business will grow.

CLOSING THE SALE

Closing the sale is the most important responsibility of a salesperson. As a contractor, closing a sale may not be the first responsibility in your job description, but without sales, you will not have a job. Like it or not, you must become sales orientated. If you just don't have the ability or inclination to make sales, hire a sales professional. One way or the other, you need sales.

The easiest way to close a sale is to ask the customer for the sale. You might be surprised at how many contractors are afraid to ask for a "yes." If you aren't willing to ask the customer to commit to your offer, you won't get many jobs. Customers expect you to ask for the sale, so you may as well accommodate them.

Not all closings are easy. I have spent hours closing sales. I have been asked to leave as many as three times and still made the sale. Over the years I have developed many sales skills. I learned how to sell by reading how-to books and through on-the-job experience. I don't think I could have been as successful without both of these learning opportunities. The books gave me ideas and formulas, the real-world experience allowed me to refine what I read and to develop my own style.

Back when I was keeping track, I had a closing ratio of 65 percent. That means that I was closing more than half of the jobs I went out to estimate. You may be thinking I was giving the jobs away, but I wasn't. Many times my prices were higher than other bids, but I won the jobs with my sales skills. You can too.

Until you learn to close sales, your business will not work at its maximum production. Take some time to study books on sales techniques and don't be afraid to try them out. Even if you are not a master at sales tactics, you will be ahead of most of your competition. Never underestimate the power of marketing, advertising, and sales skills.

SATISFY YOUR CUSTOMERS

When you are learning how to keep your customers satisfied, you must look at each customer individually. There will, of course, be similarities between customers, but each person will have at least a slightly different opinion of what is required of you. There are some basic principles to follow when working with customers:

- Keep your promises.
- Return phone calls promptly.
- Maintain an open and honest relationship.
- Be punctual.
- Listen to the requests of your customers.
- Do good work.
- Don't overcharge consumers
- Stand behind your work.
- Be professional at all times.
- Don't take your customers for granted.
- Put priority on warranty work.
- When feasible, give customers what they want.
- If possible, give customers more than they expect.

If you follow the basic rules of customer satisfaction, you should have a high ratio of happy patrons. There will, of course, be some customers you can't satisfy. Some of these people won't know themselves what it will take to make them happy. When you run into this type of client, grin and bear it.

Even if a customer is being irrational, go to extremes to please him. One angry customer will spread more word-of-mouth advertising than 10 satisfied customers, and you don't want the type of publicity a disgruntled customer will give you.

If you reach a point where you can't deal with an unreasonable person, end the connection as quickly and professionally as you can. Avoid name-calling and arguments. If the customer is dead wrong, defend your position. But if the circumstances are questionable, cut your losses and get out of the game. Avoiding conflicts and striving for customer satisfaction will do you more good than standing on a soapbox, screaming to the world what a jerk someone is.

LEARN WHEN TO GIVE AND TAKE

If you make a habit of being too considerate and giving, some of the people in this world will take advantage of you. If you are cold and take a stout stand against giving in, you will lose business. To have a harmonious business, you must learn to blend give-and-take into a masterful mix. For success in business, you have to learn the fine art of flexibility.

People go into business to be their own boss, but as a business owner you have more bosses than ever. Every customer you serve is your boss. Being in business is not the easy ride some people think it is. Owning a business is hard work, and sometimes you have to do things you don't want to. Learning to compromise, especially on money, can be a sobering experience. But if you are going to stay in business, you are going to have to make compromises. The key is learning the difference between compromise and giving in.

When you disagree with a customer, listening may be your best move. If you listen carefully to your customer's complaint, you can often find a simple solution to the problem. If, on the other hand, you do all the talking, you will probably only worsen the situation.

Becoming a good listener is one of the best ways to keep your customers happy. Once you have heard the customer out, propose a reasonable plan to resolve the dispute. You may not come to terms on the first attempt, but if you follow this procedure, you have a good chance of finding common ground.

ESTABLISH COMMUNICATION CHANNELS

Your first contact with customers will often come over the phone lines. In this first contact, you will not be able to read body language and facial expressions. What is said and the way it is said is all either of you will have to judge each other. For this reason, it is important to speak clearly and project a cheerful attitude.

Once you get past the phone conversation, a face-to-face meeting will likely occur. During this meeting, pay attention to the words and actions of your prospective customers. Also, be selective in your conversation. If you use four-letter words on the job site, keep them out of the meeting. If you smoke, refrain from your habit during the meeting. If you watch and listen to the people you are meeting, you will learn how to handle the potential clients.

When you get to the estimate and contract stage, keep your thoughts concise and in writing. Once the written documents are drawn up, go over them with your customers. If the customers want to have the documents reviewed by an attorney, by all means, let them. You want to establish a comfort level for the customer. By keeping your contracts clear and easy to read, you will keep your customers happy. Clear communications are essential to good business.

CALM A DISGRUNTLED CUSTOMER

A customer can be dissatisfied without being physically upset. In fact, the calm customer who is displeased can be more difficult to work with than the customer who is shouting in rage. When you have an unhappy customer on your hands, you have to find a remedy that will appease the client. If you are able to carry on a normal conversation with the consumer, resolving the problem shouldn't be too difficult.

Ask your customer why he is dissatisfied. Before making a rebuttal, consider the other person's position. Does he have a legitimate gripe? If he does, take action to rectify the situation. If you disagree with the customer's opinion, discuss the problem in more depth.

Before you start a debate, put the customer at ease. Assure him that you are willing to be reasonable, but that you need more facts to understand his position fully. By starting this way, the customer should remain calm and businesslike. On the other hand, if you aggressively open your defense, the situation may escalate to a tense and unpleasant shouting match.

It is usually best to attempt to settle disputes in a relaxed atmosphere. If your crews are banging hammers and buzzing saws, ask the customer if the two of you can find a more suitable place to discuss your differences. This accomplishes two goals—you get the customer in a congenial setting, and your workers don't witness the disagreement.

After relocating, ask the client to repeat his grievance. Pay close attention, and see if the story remains the same. If the customer comes up with additional complaints or a great variance from the initial comments, you may have additional trouble. This behavior might indicate a person that is going to be hard to please. However, if the complaint is essentially the same as it was when you first heard it, you have a good chance of ending your discussion in concurrence.

Think before you speak. If you are good at thinking on your feet, you will do better than people who must meditate before coming to a conclusion. Your customer is going to expect answers now, not next week. Once you know what you want to say, say it sincerely and with conviction. Let the customer know you believe strongly in your position, but that you are willing to compromise.

You may find that you and your customer will exchange several opinions and offers before you reach an amicable decision. Once the two of you agree on a plan, put the plan in writing. By writing a change order for the compromise, you seal the deal and reduce the risk of having to negotiate it further at a later date.

BUILD A REFERENCE LIST FROM EXISTING CUSTOMERS

Ask your customers to give you a letter of reference or to complete a performance-rating card and you will compile a valuable stack of ready references. Asking customers for the names and phone numbers of friends or relatives is another way to get new business.

For example, assume you are nearing completion on a remodeling job. You ask the customer for the names and phone numbers of his neighbors. That evening, you call the neighbors and introduce yourself as the contractor working on Mr. Smith's remodeling project. You explain to the neighbors that the Smith job is nearing completion and that you wanted to see if they had any work that needed to be done before you pulled your crews and equipment out of the area. Further explain how you can offer special pricing and discounts for any work they want done now, since your people and equipment are already working in the neighborhood.

By knowing the neighbor's name, you have a good chance of starting a conversation before they hang up on you. By giving Mr. Smith's name as a reference, you have a good chance of keeping the conversation going. Your offer to discount your services sounds viable, and you may be surprised at how many neighbors will take you up on your offer.

AVOID ON-THE-JOB DECISIONS

When you make snap decisions, you are likely to make mistakes. These mistakes could run the gamut from upsetting the homeowner to causing extra work for yourself. When possible, give yourself time to consider your decisions before you make them. Of course, there will be times when you have to shoot from the hip, but avoid quick decisions when you can. Let's consider how making decisions too quickly can result in disaster.

In our first example, consider you have to make an on-site adjustment to accommodate the installation of a plumbing stack. Your plumber needs to get a 3-inch pipe into the attic you are converting to living space. The problem is this: there is no feasible way to get the pipe into the attic without altering the downstairs living space.

The plumber calls you to the job to make a decision on which way you want the pipe routed to the attic. The homeowners are at work. Your options include: opening an existing wall, building a chase for the pipe to run up, or using the corner of a closet to get the pipe into the attic.

The plumber wants an immediate decision on which way to go with the pipe. Since the plumber is working on a contract basis, all the time lost in making the decision is coming out of the plumbing profits. What are you going to do? Should you call the homeowners and get their opinions, or should you make your own decision? You should call the homeowners, but what will you do when they ask your opinion?

I would opt for installing the pipe in a closet, but the homeowner may not agree. Don't forget to consider the cost of this change in plans. Since you gave the customers a contract price, you cannot charge them for the extra expense. Your plumber may be willing to absorb the cost of lost time, but it is unlikely the plumber will cover the other costs involved.

If you cut open a wall or build a chase, your expenses will be more than if you hide the pipe in a closet. Carpentry work, drywall work, and paint work will be required to build a chase or repair a finished wall. By running the pipe in the closet, a few trim boards can form a box that will hide the pipe. You can see why I would favor putting the pipe in the closet; it's the least expensive way to get out of the loss.

If you haven't considered all of your options when the homeowners ask your opinion, you may come off looking inexperienced. If you choose the wrong option, the homeowners may resent your advice by the end of the job. Picking the wrong pipe placement could cost you additional money. This may not be a job-shattering decision to make, but it can have undesirable consequences.

In this next example, you are having problems with one of your subcontractors. The sub has started the job and is about half done. However, the subcontractor is not performing to the satisfaction of the customer. The customer calls and tells you to replace the existing sub with someone who will do a better job. What are you going to say?

If you refuse to eject the subcontractor from the job, the customer is going to be angry. If you do fire the subcontractor, the sub is going to be upset. At first glance, it makes sense to keep the customer happy, even if it means upsetting the independent contractor. But, there is more to this decision than meets the eye.

Unless your contract with the subcontractor gives you the right to remove the sub if the work is not acceptable, you have a serious problem. Without a clause giving you special powers, preventing the subcontractor from fulfilling the contract will be a breach of the contract. Under these conditions, you may have to pay the sub in full, even if the job is only partially done. If you refuse to pay the contracted amount, the sub might place a mechanic's lien against the property. If this happens, you can be sure the homeowner will go through the roof.

This type of problem is common and delicate. To avoid this situation, you need strong language in your subcontractor agreement. Without an ejection clause, you can be rendered nearly helpless. However, if you find yourself in this bind, make your decision carefully. If you make the wrong choice, you could find yourself surrounded by people in vicious moods. Even if your on-the-spot decision seems trivial, think it through. It is easy for little problems to balloon into big ones.

LIEN RIGHTS AND WAIVERS

As a contractor, you will probably be asked to sign lien waivers and you will most likely ask subcontractors to sign lien waivers. When a job is being financed, the lender will often require lien waivers to be signed for every cash disbursement. When the person being paid signs a lien waiver, that person gives up rights that may exist to lien the property for the labor or materials. This protects the property owner and the lender.

Lien rights and waivers can have a significant impact on your business. These factors can work for you or against you. A lien right is the right to place a lien against property to which you have supplied labor or material and not been paid. A lien waiver is a legal document that when signed, relinquishes your lien rights.

There are two types of liens that generally come into play with contractors. A mechanic's lien is a lien that may be placed by people who have not been paid for labor provided on a job. A materialman's lien is a lien that can be levied by suppliers that have not been paid for materials supplied to a job.

Lien-right laws vary from jurisdiction to jurisdiction, but they exist to protect workers and suppliers. To gain a full insight into the lien rights available to you, consult a local attorney.

Some contractors are approved for short-form lien waivers and others must use long-form lien waivers (FIGS. 14-1 and 14-2). A short-form lien waiver is a form that only the general contractor signs. In signing a short-form waiver, the contractor is swearing that all subcontractors have been paid for work done to a certain point. The contractors that are approved for short-form lien waivers have usually been in business for awhile and normally have strong company assets.

Property owners and lenders are taking a bigger risk when they allow contractors to sign short-form waivers. If the general contractor signs the waiver, but has not paid the subcontractors or suppliers, the unpaid parties may still lien the property. The general contractor will be responsible for having the lien removed, but there is some risk that the general contractor will not have the funds to settle the issue and have the lien removed.

Long-form lien waivers are more time-consuming for contractors, but the property owners and lenders are in a much safer position. With a long-form lien waiver, anyone providing labor or materials for a job must sign the lien waiver at the time of payment. This way, no one can say they weren't paid; their signature will be on the lien waiver.

As a general contractor, it is good business to have subcontractors and suppliers sign lien waivers, even if a lender or property owner is not requiring the paperwork (FIG. 14-3). Have your vendors sign off on the waivers when they are paid and you will be assured that liens will not be placed against the properties you service. This is just one more paperwork step that can help you avoid conflicts and trouble.

SOLIDIFY AGREED-UPON PLANS AND SPECIFICATIONS

Once you and your customer agree on a set of plans and specifications, memorialize the agreement—have your customers sign them. Date the documents and make notations that the documents are the final and working plans and specifications. Further note that no changes will be made to the documents unless all parties agree to the changes on a written change order. When your customers sign below these notes, you have a solid set of plans and specifications. This procedure will eliminate the possibility of the customers coming to you later and saying that the job is not in compliance with the plans and specs.

The signed agreements are hard to value. Having clear contracts and supporting documents will keep you out of ambiguous arguments. Your signed documentation will prove who is in the right.

Long-Form Lien Waiver

Customer name: ______________________________

Customer address: ______________________________

Customer city/state/zip: ______________________________

Customer phone number: ______________________________

Job location: ______________________________

Date: ______________________________

Type of work: ______________________________

The vendor acknowledges receipt of all payments stated below. These payments are in compliance with the written contract between the vendor and the customer. The vendor hereby states that payment for all work done to this date has been paid in full.

The vendor releases and relinquishes any and all rights available to said vendor to place a mechanic or materialman lien against the subject property for the described work. Both parties agree that all work performed to date has been paid for, in full and in compliance with their written contract.

The undersigned vendor releases the customer and the customer's property from any liability for non-payment of material or services extended through this date. The undersigned contractor has read this entire agreement and understands the agreement.

Vendor Name	*Signature of Co. Rep.*	*Signature Date*	*Service Performed*	*Date Paid*	*Amount Paid*
Plumber (Rough-in)					
Plumber (Final)					
Electrician (Rough-in)					
Electrician (Final)					
Supplier (Framing lumber)					

*This list should include all contractors and suppliers. All vendors are listed on the same lien waiver, and sign next to their trade name for each service rendered, at the time of payment.

14-1 Long-form lien waiver.

Short-Form Lien Waiver

Customer name: ____________________

Customer address: ____________________

Customer city/state/zip: ____________________

Customer phone number: ____________________

Job location: ____________________

Date: ____________________

Type of work: ____________________

Contractor: ____________________

Contractor address: ____________________

Subcontractor: ____________________

Subcontractor address: ____________________

Description of work completed to date: ____________________

____

____

____

Payments received to date: ____________________

Payment received on this date: ____________________

Total amount paid, including this payment: ____________________

The contractor/subcontractor signing below acknowledges receipt of all payments stated above. These payments are in compliance with the written contract between the parties above. The contractor/subcontractor signing below hereby states payment for all work done to this date has been paid in full.

The contractor/subcontractor signing below releases and relinquishes any and all rights available to place a mechanic or materialman lien against the subject property for the above described work. All parties agree that all work performed to date has been paid for in full and in compliance with their written contract.

The undersigned contractor/subcontractor releases the general contractor/customer from any liability for non-payment of material or services extended through this date. The undersigned contractor/subcontractor has read this entire agreement and understands the agreement.

____

Contractor/Subcontractor Date

14-2 Short-form lien waver.

Certificate of Subcontractor Completion Acceptance

Contractor: ______________________________

Subcontractor: ______________________________

Job name: ______________________________

Job location: ______________________________

Job description: ______________________________

____

____

____

____

____

Date of completion: ______________________________

Date of final inspection by contractor: ______________________________

Date of code compliance inspection & approval: ______________________________

Defects found in material or workmanship: ______________________________

____

____

____

____

Acknowledgment

Contractor acknowledges the completion of all contracted work and accepts all workmanship and materials as being satisfactory. Upon signing this certificate, the contractor releases the subcontractor from any responsibility for additional work, except warranty work. Warranty work will be performed for a period of one year from the date of completion. Warranty work will include the repair of any material or workmanship defects occurring between now and the end of the warranty period. All existing workmanship and materials are acceptable to the contractor and payment will be made, in full, according to the payment schedule in the contract, between the two parties.

______________________________ ______________________________

Contractor Date Subcontractor Date

14-3 Certificate of subcontractor completion acceptance.

15
Bids open new markets

Building business clientele is one of your most important jobs as a business owner. Learning how to win bids is one of the most effective ways to build your business clientele. How many times have you bid a job and never heard back from the potential customer? Many contractors never figure out how to win bids successfully. This chapter is going to show you how to win bids and build up your business.

REACH OUT FOR A NEW CUSTOMER BASE

Reach out for a new customer base through bid sheets. Bid sheets are open to all reputable contractors. When a job is placed on a bid sheet, somebody is going to get the job. The jobs put out to formal bids are almost always done. Unlike common residential estimates, where potential customers change their minds, formal bid sheets are rarely changed. This type of work is very competitive and the percentage of profit is usually low, but bid work can pay the bills.

Where do you get bid sheets?

Bid sheets can be obtained by responding to public notices in newspapers and by subscribing to services that provide bid information. If you watch the classified section of major newspapers, you will see advertisements for jobs going out to bid. You can receive bid packages by responding to these advertisements. Normally, you will get a set of plans, specifications, bid documents, bid instructions, and other needed information. Bid packages can be simple or complicated.

What is a bid sheet?

A bid sheet is a formal request for price quotes (FIG. 15-1). There is a difference between a bid sheet and a bid package. The bid sheet will give a brief description of the work available. A bid package gives complete details of what will be expected from bidders. Most contractors start with a bid sheet and if they find a job of interest, order a bid package. Bid sheets are usually provided free of charge. Bid packages often require either a deposit or a nonrefundable fee.

Bid Request

Customer name: ______________________________

Customer address: ______________________________

Customer city/state/zip: ______________________________

Customer phone number: ______________________________

Job location: ______________________________

Plans & specifications dated: ______________________________

Bid requested from: ______________________________

Type of work: ______________________________

Description of material to be quoted: ______________________________

All quotes to be based on attached plans and specifications. No substitutions allowed without written consent of customer.

Please provide quoted prices for the following: ______________________________

____

All labor, materials, permits, and related fees to complete plumbing as per attached plans and specifications.

All bids must be submitted by: ______________________________

15-1 Bid request form.

Bidder agencies

Bidder agencies are businesses that provide listings of bid opportunities. These listings are normally published in a newsletter form. The bid reports are generally delivered to contractors on a weekly basis. Each bid report may contain from 5 to 50 jobs. These publications are an excellent way to get leads on all types of jobs.

What type of jobs are on bid sheets?

All types of jobs appear on bid sheets. The work can range from small residential to large commercial jobs. The majority of the jobs are commercial. The size of the jobs ranges from a few thousand on up to millions of dollars.

Government bid sheets

Government bid sheets are another place to find an abundance of work. Like other bid sheets, government bid sheets give a synopsis of the job description and provide information on how to obtain more details. Government jobs can range from replacing a dozen lavatory faucets to building a commissary.

Government jobs are a safe bet for getting your money. The money may be slow in coming, but it will come. The paperwork involved with government jobs can be excessive. If you are not willing to deal with mountains of paperwork, stay away from government bids.

ARE YOU BONDABLE?

Many jobs found on bid sheets require contractors to be bonded. Bonds are obtained from bonding companies and insurance companies. Before you try bidding jobs that require bonding, check to see if you are bondable. The requirements for being bonded vary. Check in your local phone book for an agency that does bonding, and call to inquire about the requirements.

PERFORMANCE AND SECURITY BID BONDS

Performance and security bid bonds are a necessity for many major jobs. If you order a bid sheet or package, you will almost certainly see that a bond is required. Some listings on bid sheets may not require a bond. It is common for bid requirements to be tied to the anticipated cost of the job. The bigger the job, the more likely it is a bond will be required.

Bonds are required to ensure the success of the job. The people offering the work want to be sure that the job will be done right and that it will be completed. When the person or firm issuing the work requires a bond, they know there is a degree of safety.

It is very difficult for some new businesses to obtain a bond. If the new company doesn't have strong assets or a good track record, getting a bond is tough. Unfortunately, a bond can be a hurdle you can't get over until you don't care whether you have it or not.

There are three basic types of bonds: bid bonds, performance bonds, and payment bonds. Each type of bond serves a different purpose. A bid bond is put up to assure the person receiving bids that the bidding contractor will honor the bid if a contract is offered.

Performance bonds prevent contractors from abandoning a job and leaving the customer in dire financial straits. If a contractor reneges on completing the job, the customer may hold the performance bond for financial damages.

Payment bonds are used to guarantee payment to all subcontractors and suppliers used by a contractor. These bonds eliminate the risk of mechanic and materialman liens being filed against the property where work is being done.

When you put up a bond, the value of the bond is at risk. If you default on your contract, you lose your bond to the person that contracted with you for the job. Since many people use the equity in their home for collateral to get a bond, they could lose their house. Bonds are serious business. If you can get a bond, you have an advantage in the business world. Talk with local companies that issue bonds to see if you qualify for bonding.

BIG JOBS EQUAL BIG RISKS?

Are big jobs surrounded by big risks? You bet they are. There are risks in all jobs, but big jobs do carry big risks. Should you shy away from big jobs? Maybe, but if you go into the deal with the right knowledge and paperwork, you should survive and possibly prosper.

Cash-flow problems

Cash-flow problems are frequently present with contractors doing big jobs. Unlike small residential jobs, big jobs don't generally allow for contractors to receive cash deposits. If you tackle these jobs, you will have to work with your own money and credit. For a new business, the money needed to get to a draw disbursement in big jobs can be the undoing of the company. It's wonderful to think of signing a million-dollar job in your first year of business, but that job could put your business into the bankruptcy court.

Before you dive into deep water, make sure you can get to the other side. Some lenders will allow you to use your contract as security for a loan, but don't bet your business on it. If you want to take on a big job, first get your finances in order.

Slow pay

Slow pay can be another problem with big jobs. Large jobs are notorious for slow pay. It's not that you won't get paid, but you may not get paid in time to keep your business going. New businesses are especially vulnerable to slow pay. When you move into the big leagues, be prepared to hold your financial breath for awhile. The check you thought would come last month might not show up for another 90 days.

No pay

Slow pay is bad, but no pay is worse. Just as new contractors can get in trouble with large jobs, developers and general contractors can get into financial difficulty with big jobs. Most of the people spearheading big jobs don't intend to stick their subcontractors, but sometimes they do. When the top dogs get in over their heads, they can't pay their bills.

If the subcontractors don't get paid, suppliers don't get paid. The ripple effect continues. Everyone involved with the project will lose. Some will lose more than others. Generally, when these big jobs go bad the banks or lenders financing the whole job will foreclose on the property as they normally hold a first mortgage on the property.

For subcontractors, filing mechanic liens is the best course of action. If a contractor hasn't been paid for labor or materials, a mechanics lien can usually be levied against the property where the labor or materials were invested. If you have to file a lien, make sure you do it right. There are rules you must follow when filing and perfecting a lien. You can file your own liens, but I recommend working with an attorney on all legal matters.

Even after you file and perfect your lien, you may not get your money. If you get any money, it will likely be a settlement for a reduced amount. You can never quite get the taste of sour jobs out of your mouth.

Completion dates

Completion dates can also wreak havoc with the inexperienced contractor. Big jobs often include a time-is-of-the-essence clause. Along with this clause is usually a penalty fee that must be paid if the job is not finished on time. The penalty is normally based on a daily fee. For example, you may have to pay $200 per day for every day the job runs past the deadline.

Penalty fees and the possible loss of your bond can ruin your business. Contractors with limited experience in big jobs are often unprepared to project solid completion dates. Don't sign a contract with a completion date you are not sure you can meet.

ELIMINATE YOUR COMPETITION

When you learn how to eliminate your competition in the bid process, you are on your way to a successful business. There is no shortage of competition in most contracting fields. There are, however, often work shortages. With the combination of limited work and unlimited competition, a new business owner, or any business owner for that matter, can get discouraged. But don't, there are ways to thin out the competition.

Unless you have a track record and are well known, money will talk. Low prices are what most decision makers look for on bid sheets. Being the low bidder can get you the job, but make you wish later that you had never seen the job. Don't bid a job too low. It doesn't do you any good to have work if you're not making money.

How can you improve your odds in mass bidding? If you can get bonded, you will have an edge. A lot of bidders can't get bonded. This fact alone can be enough to cull the competition. When you prepare your bid package for submission, be meticulous. All you will have going for you will be your bid package. If you want the job, prepare a professional bid packet.

If you are dealing with in-person bids, follow the guidelines found throughout this book. The basic keys include:

- dress appropriately
- drive the right vehicle
- be professional
- be friendly
- get the customer's confidence
- produce photos of your work
- show off your letters of reference
- give your bid presentation in person
- follow up on all your bids

16
Develop your public image

How much is your public image worth? The type of business image you present might mean the difference between success and failure. Does it really matter if you don't have a logo? Yes, people quickly learn to associate a logo with its owner. Logos can do a lot for you in all your display advertising.

How important is the name you choose for your company? Give careful consideration to your company name. Some names are easier to remember than others. A name should conjure up the image you want for your business. If you intend to sell your business in the future, your company name should not be too personal. The new owner may not like owning a business with your personal name as part of the company name.

A company image can affect the type of customers your business attracts and the rates that you may charge for services. Your business advantage can come from a high profile in community organizations. How you shape your business image may set you apart from the crowd and help eliminate normally heavy competition.

The value of a public image is hard to determine. While it may be almost impossible to set a monetary value for your company image, it is easy to see how a bad image will hurt your business. Your image has many facets. Your tools, trucks, signs, advertising, and uniforms will all have a bearing on your corporate image. This chapter is going to detail how these and other factors work to make or break your business.

BUILD POSITIVE PUBLIC PERCEPTIONS

How the public perceives your business is half your battle. If you give your customers the impression of a successful business, you probably will be successful. On the other hand, if you don't take an active interest in building a strong public image, your business may sink into obscurity.

Take your truck as an example. If you had a contractor come to your house to give an estimate, how would you feel if he arrived in an old, battered pick-up truck, with bald tires and a license plate hanging from baling wire? Would you rather do business with this individual or a person pulling up in a late model, clean van that had the company name professionally lettered on the side? Which truck points to the most company success and stability? Most people would prefer to do business with a company that gives the appearance of being financially sound. This doesn't mean you have to have flashy new trucks, but they should be modern and clean.

Is it important to have your company name on your business vehicles? You bet it is. The more often people see your trucks around town, the more they will remember your name and develop a sense of confidence in you. It is acceptable to use magnetic signs or professional lettering, but don't letter the truck with stick-on letters in a haphazard way. Remember, you are putting your company name out there for all the world to see.

Designing your ad for the phone directory is another major step. As people flip through the pages of the directory, a handsome ad will stop them in their tracks. An eye-appealing ad can get you business that would otherwise be lost to competitors.

While we are talking about phone directories, let's not forget about phone manners. Telephones often provide the first personal link between your business and potential customers. You can't afford to lose customers at the inquiry stage. Don't allow small children to answer your business phone. A rude answering service is a sure way to lose potential business. Answering machines may cost you some business, but they have become an acceptable way of doing business. Don't try to get too cute on your machine's answering message. Your customers expect a professional response. If you put a tape in your answering machine that is ridiculous or offending, would-be customers are sure to hang up.

Any professional salesperson will tell you that to be successful, you must always be in a selling mode. It doesn't matter where you are or what you're doing, you must be ready always to cultivate sales. Compare this fact to your public image. Your image is being made and presented every day and in every way. You can't afford to let your business image slip.

If your company image is strong enough, customers will come to you. They will see your trucks, job signs and ads, and call you. When a customer calls a contractor, they are usually serious about having work done. Whether your company does the work, or one of your competitors gets the job, will depend upon many factors. Your company image is one of those factors. By building and presenting the proper image, you are halfway to the closing of the sale.

SEPARATE YOURSELF FROM THE CROWD

The name and logo you choose will be with the company for many years. In order to make your business better than average, you must set yourself apart from the crowd. Before you decide on a name or a logo, do some research and some thinking. Ask yourself several questions. For example, do you plan to sell the business in the future? If you do, pick a name that anyone can use comfortably. A name like Pioneer Plumbing can be used by anyone, but a name like Ron's Remodeling Service is a little more difficult for a new owner to accept. This is only one example; let's move on to other factors you should consider before you choose a name and logo.

Company name

Company names say a lot about the business they represent. For example, High-Tech Heating Contractors might be a good name for a company specializing in new heating technologies and systems. Solar Systems Unlimited might be a good name for a company that deals with solar heating systems. Authentic Custom Capes would make a good name for a builder that specializes in building period-model Cape Cods. What image will a name like Jim's Plumbing present? A better choice might be, Jim's 24-Hour Plumbing Service. Now the name tells customers that Jim is there for them, 24 hours a day. See how a name can influence the public's perception of your business?

Your company name should be one that you like, but it should also work for you. If you can imply something about your business in your name, you have an automatic advantage.

If you were scanning through the newspaper and noticed a company name like Tanglewood Enterprises, what association would you make? This name could apply to any business. On the other hand, if you saw a name like Deck Masters, Inc., you would probably associate the name with decks. If the name was White Lightning Electrical Services, you would think of an electrical company. A name like Jim's 24-Hour Plumbing Services leaves a clear impression of a company that offers 24-hour plumbing service. The more you can tie the name of your business to the type of business you are in, the better off you will be.

It also helps when you choose a name to find words that flow together smoothly. How does Pioneer Plumbing sound to you? Both words start with a "P," and the words work well together. A name like Ron's Remodeling sounds good; so does a name like Mike's Masonry. In contrast, a name like Englewood Heating and Air Conditioning isn't bad, but also isn't good. A name like Septic Suckers flows well and might be fitting for a company that pumps out septic tanks, but the name may not generate business. Although, the more I think about it, the more I like the name. It might just generate enough talk around town to keep your phone ringing.

Logo

A logo is as important as your company name. Your logo, the symbol you adopt to represent your company, plays an important role in marketing and advertising (FIG. 16-1). While people may not remember a specific ad or even a company name, they are likely to remember a distinctive logo. If you

16-1 Example of company logo.

put your mind to it, I'll bet you can come up with at least 10 logos that stick in your mind.

Major corporations know the marketing value of logos. They invest considerable time and money in an effort to come up with just the right symbol. While you may not go to the expense of hiring an ad agency or graphic artist to design a custom logo for your business, it will still pay off in the future to use a logo in your advertising. Like slogans and jingles, logos are often much easier to remember than company names.

Your logo doesn't have to be complex. In fact, it might be nothing more than the initials of your company name. Then again, you may have a very complex logo, one that incorporates an image of what your business does. One of my favorite logos was used by a real estate company. The logo was a depiction of Noah's ark, complete with animals. In the ad featuring the ark were the words, "Looking for land?" Their business was selling land, and the ark logo was humorous and fitting.

If your mind is not at its best when it comes to creative images and marketing, you may want to consult a specialist in the field. Choosing the proper name and logo is important enough to warrant investing some time and money.

Company colors

Choose your company colors with care. Company colors are one way to attract attention and become known all over town. Colors affect how people think, the mood they are in, and what they do. How seriously would you take a plumber who pulled up in a pink van with flowers painted on it? The color and decoration of the plumber's van has no bearing on the technical ability of the plumber, but it does cast an immediate impression.

The color of your trucks may be dictated by the color of the truck you presently own. It is more impressive to see a fleet of trucks that are uniform in color and design than it is to see a parade of trucks that include various makes and colors. A unified fleet gives a better impression.

The color you select for your truck lettering and job-site signs is also important. If your truck is dark blue, white letters will show up better than black letters. If the truck is white, black or blue letters would be fine. For job signs, it is important to pick a background color and a letter color that contrast well. You want the sign to be easy to read. When you talk with your sign painter or dealer, review examples of how different colors work together.

Slogan

Slogans are often remembered when company names are not. If you advertise on radio or television, slogans are especially important. Since radio and television provide audible advertising, a catchy slogan can make its mark and be remembered. When advertising in newspapers or other print ads, slogans associate key words with your company.

Thinking of a slogan for your business might take a while, but it's worth the effort. If you need inspiration, look around at other successful companies. Examine their slogans to gain ideas for yours, but never use someone else's slogan.

ORGANIZATION MEMBERSHIPS GENERATE LEADS

Joining clubs and organizations is an excellent way to generate sales leads. Whether you like it or not, as a business owner you must also be a salesperson. When you join local clubs and community organizations you meet people. These people are all potential customers.

Become visible in your community and your business will have a better chance of survival. If you support local functions, children's sports teams and the like, you become known. You can use the local opportunities to build your business image. When citizens see your company name on the uniforms of the local kids' baseball team, they remember you. Further, they respect you for supporting the children of the community. You can take this type of approach to almost any level.

IMAGE AND YOUR FEE SCHEDULE

Image may not be everything in business, but it is a big part of your success. The public has read all the horror stories of ripoffs and contractor con artists. Unfortunately, much of what has been printed about unscrupulous contractors is true, and the public does have a right to be concerned. With growing customer awareness, image is more important than ever before.

If you dress neatly, in casual clothing, and drive a respectable, but not lavish vehicle, your odds of appealing to the masses improve. Your vehicle should look professional and successful, without reeking of money. For me, this combination has always worked best. By wearing clothes that allow you to crawl into the attic or under the sink, you give the impression of someone who knows the contracting business.

The same basic principles apply to your office. If your office is little more than a hole in the wall with an answering machine, an old desk, and two chairs, people will be concerned about the financial stability of your business. However, if your office is staffed with several people, decorated in expensive art and furnishings, customers will assume your prices are too high. It generally works best to hit a happy medium with your office arrangements.

The image you present has a direct affect on your fees. If you convince potential customers you are a professional, your customers will be willing to pay professional fees. Extend your image by making the customers feel safe in their business dealings with you and you have leverage for higher fees. When you become a specialist, you can demand even higher fees. Think about it, who gets a higher hourly rate, your family doctor or a heart specialist? When you convince the customer that nobody builds a better garage than you do, you are building a case for higher fees.

For years I specialized in kitchen and bath remodeling. My crews did nothing but kitchen and bathroom remodeling. When you do the same type of work day in and day out, you get good at it. With my experience in this specialized field of remodeling, I anticipated problems and found solutions that most of my competitors did not. This specialized experience made me more valuable to consumers. I could snake a 2-inch vent pipe up the wall from their kitchen to their attic without cutting the wall open. I could predict with great accuracy how long it would take to break up and patch the concrete floor for a basement bath. In general, I became known as a competent professional in a specialized field, and I could name my own price, within reason. You can do the same thing.

If you have a special skill and can show the consumer why you are more valuable than your competitors, the consumer is very likely to pay a little extra for your expertise. Building a solid image as a professional that specializes in a certain field has its advantages.

IT IS DIFFICULT TO CHANGE AN IMAGE

Once you cast an image it is difficult to change. If you are already running an established business, don't be discouraged by this comment. While it is harder to change an established image than it is to create a new one, it is not impossible. Work to change your image. With enough time, effort, and money, you can make a difference in your company image.

Say you want to change the name of your existing company. Send out letters to all of your customers advising them of your new company name. Explain that due to growth and expansion you are changing the name of the company to reflect your growth. Impress upon the existing customers that the company has not been sold and is not under new management, unless you have a bad image to overcome, then the new-management announcement might be a good idea.

Place new advertisements with your new company name and logo. Build new business under your new name and convince past customers to follow you in your expansion efforts. By taking this approach, you get a new public image without losing the bulk of your past customers.

While this approach will work, it is best to create a good image when you begin the business. It is always easier to do the job right the first time than it is to go back and correct mistakes.

17 Marketing and advertising strategies

Marketing and advertising strategies may well be the most important lesson for new business owners to learn. While it is true that advertising alone will not make a business successful, it is a critical element when building a thriving business.

To get the opportunity to make sales, most businesses must advertise. Without advertising, the average business will not get many customer inquiries. If no one knows your business exists, how will they contact you for service? Since public exposure is paramount to the success of your business, so is a strong marketing plan and effective advertising.

Too many contractors fail to see the importance of marketing and advertising. For some reason, many contractors think the public will seek them out. Let me repeat myself, if the public doesn't know you exist, they can't very well seek you out. Regardless of how good you are at what you do, you won't get much work unless people are aware of your services.

To get busy and stay busy, you need regular sales. Marketing and advertising can provide you with sales leads. It will be up to you or your salespeople to convert the leads into closed sales, but you must start by getting prospects who want what you have to offer. Advertising is the most effective way to quickly generate lends.

MARKETING IS PIVOTAL POINT FOR YOUR BUSINESS

Marketing is the active process during which you react to the business climate. When you track your advertising results, design your ads, develop sales strategies, and define your target market, you are performing marketing skills.

Marketing demands an extension of the normal senses. You must be able to read between the lines and determine what the buying public wants. There are many books available on marketing. Professional seminars teach marketing techniques. Many community colleges offer courses in marketing. With enough effort and self study, you can become an efficient marketer. If you want to have a business with a long life, develop your marketing skills.

SHOULD YOU ADVERTISE?

Advertising is the act of putting your message in front of consumers. You can advertise in newspapers, on radio, on television, by direct mail, or in many other ways.

Advertising is expensive, but it is also a necessary part of doing business. If you don't spend money on advertising, the public is not going to spend money with your business. The contracting field is filled with business owners who are aggressive. These aggressive owners advertise regularly. If you don't put your name in front of people, you will be run over by your competition.

Advertising in the local phone directory will generate customer inquiries and provide credibility for your company. Ads in the local newspaper can result in quick responses. Door-to-door pamphlets and fliers can produce satisfactory results. Radio and television ads can be very effective. Putting a slide-in ad on the video boxes at the local video rental store can give you a lot of exposure. The list of possible places to advertise is limited only by your imagination. However, some advertising media are better than others. Let's take a close-up look at some specific examples.

Phone directory

The phone directory is an excellent place to advertise your company name. Being listed in the phone book adds credibility to your company. Whether you are merely listed in a line listing or have a full-page display ad, you should get your company name in the phone book as soon as possible.

The size of your ad in the directory should be determined by the results you hope to achieve. Large display ads are expensive, and they may not pay for themselves in your line of work. If you are a plumbing company that offers emergency repair service, a large ad is beneficial. People with a basement that is flooding or a toilet that is running over in the upstairs bath are in a hurry to find a plumber. Most phone directories place the larger ads in the front section of the category heading. Being the first plumbing company a panicked homeowner comes to can give you a real advantage.

If your business is building houses, a large display ad probably isn't necessary. When people are shopping for a builder, they are not normally in a hurry. An ad that is one column wide and an inch or two in length might pull just as many calls under these circumstances. A quick look at how your competition advertises will give you a hint as to what you should do. If all the other builders have large ads, you probably should have a large ad, too.

As I became more knowledgeable about business, marketing, and advertising, I tested the results of various directory ads. During my test marketing, I used many types of ads for my various businesses. I found that for remodeling, real estate, and plumbing, large ads worked best. When perfecting my ads for home building, property management, consulting, and photography, I did just as well with small ads as I did with large ads.

In general, I believe that if you are dealing with people on an impulse basis, such as emergency plumbing repairs, a large display ad will pay for itself. If you are dealing in high-ticket items, like building new homes, a modest ad is sufficient.

Newspaper ads

Newspaper ads provide quick results; you either get calls or you don't. As a service contractor, my experience has shown that most newspaper respondents are looking for a bargain. If you want to command high prices, I don't think newspapers are the place to advertise. But if you are new in business, the newspaper can produce customers for you quickly.

Handouts, flyers, and pamphlets

Handouts, flyers, and pamphlets are similar to newspaper advertising. These methods seem to generate calls quickly, but the callers are usually looking for a low price. Many businesses consider this form of advertising degrading. I don't know that I would agree with that opinion, but I don't think you will receive the money you are worth with these low-cost advertising methods.

Radio advertising

Radio advertising is expensive, but it is a good way to get your name into the minds of listeners. I believe the key to radio advertising is repetition. If you can't afford to sustain a regular ad on the radio, I would advise against using this form or advertising. Most people are not going to hear your ad and run to the nearest phone to call you. However, if you can budget enough money for several radio spots over a few weeks time, you will gain name recognition.

Television commercials

Television commercials can be very effective. Unlike radio, where people only hear your ad, television allows viewers to see and hear your ad. People associate television advertisers with success. With the many cable channels available, television advertising can be an affordable and effective way to get your message out to the community.

I have used ads on cable television very effectively. My ads were for home building, but I know plumbing contractors that have been successful with this type of advertising. Whether you are selling a $30 service call or a $100,000 house, television ads can increase your sales.

Direct-mail advertising

Direct-mail advertising is effective, but not always cost effective. The cost for direct-mail advertising can easily be thousands of dollars. Most direct-mail advertisers are happy if only 1 percent of the people who receive their advertising become customers. If you are selling services for less than $50, direct mail probably won't work for you. If, however, you are selling garages, room additions, heating systems, or other big ticket items, direct mail can be a successful tool for your business.

Direct mail allows you to reach a targeted market. If you want to advertise to people with incomes in excess of $50,000, you can rent a mailing list that includes just those people. If you want to mail to recent home buyers, such mailing lists are available. With direct mail, you have complete control over who sees your advertisement.

Most mailing lists cost about $75 per thousand names. Many list brokers require a minimum order of 3,000 names. The names can be supplied to you on stick-on labels. If you want to reduce your mailing costs, your local postmaster can set you up with a bulk-rate permit. To use the bulk-rate service, you must mail a minimum of 200 pieces of mail at a time. The cost for this type of mailing is much less than first-class postage, but there are one-time and annual up-front fees. Ask your local post office for full details.

Creative advertising methods

Creative advertising methods are just that: creative. You might want to rent space on a billboard. Perhaps you will cut a deal with a local restaurant to have your company highlighted on its menus. Providing uniforms for the local little league can get your name in front of a large audience. If you put your mind to it, there are infinite possibilities for creative advertising.

BUILD A CUSTOMER BASE

Without customers, you have no business. How you develop your customer base will depend, to some extent, on the type of customers you are after. Unless you have planned the opening of your business in advance, you will not have the advantage of being listed in the phone directory. While not being in the phone book hurts your business credibility and causes you to miss some business, don't think you can't enjoy a prosperous business without a directory listing.

If you want to work for general contractors, direct mail is an effective way to get your name in front of potential customers. It is an expensive form of advertising that is fast, easily targeted, and effective. Design a professional-looking promotional letter or package to capture the attention of general contractors.

You can also target homeowners with direct mail letters, but at a much higher cost. Unlike general contractors, whom you know will need the services of subcontractors, homeowners may have no need for your services. When you make a blind mailing to residences, you have to pay the numbers game. You hope that a percentage of the individuals who receive your material are interested in what you have to offer. Unfortunately, many of the homes on your mailing list will not respond. A mass mailing to homeowners, even using bulk-rate postage, may not be cost-effective.

Local newspapers are generally an inexpensive way to advertise your company name. Most papers have a section dedicated to service companies. While response from these ads will not be overwhelming, neither will the cost of the ad.

ANALYZE YOUR RATE OF RETURN

The response to your advertising will depend on the execution of your marketing plan and the effectiveness of your advertising. If you are advertising in the local newspaper, you might expect about a .001 response. In other words, if the paper has 25,000 subscribers, you might get 25 responses to your ad. This projection is aggressive. If you only get 10 calls, don't be surprised. In some cases you may not get even 10 calls. Your advertising success will depend entirely on how well you designed the ad and matched the publication to your marketing plan.

Advertising a contracting business on radio or television might seem like a waste of money. It is not uncommon for these ads not to generate calls, but that doesn't mean the ads were not effective. Television and radio advertising builds name recognition. This form of advertising works best when it is used in conjunction with some type of print advertising.

If you are running ads in the paper, distributing flyers, or doing a direct-mail campaign when the television and radio ads are on, you should see a higher response to your print advertising than you would without the radio and television ads.

Direct-mail advertising often provides fast results. A 1 percent response on direct-mail advertising is generally considered good. For example, if you mail to 1,000 houses, you should be happy if you get 10 responses. If you target your direct-mail market to reach specific demographics, you should get a better rate of return.

Demographics are statistics that tell you facts about the names on your mailing list. You can rent a mailing list that is comprised of specific age groups, incomes, and so forth. These statistics can make a big difference in the effectiveness of your advertising.

Determining the effectiveness of advertising is a task all serious business owners must undertake. To learn what ads are paying for themselves, you need to know which ads are generating buying customers. Some ads generate a high volume of inquiries, but don't result in many sales. Other ads produce less curiosity calls and more buying customers. You need to track the results of your advertising. Without knowing which ads and advertising mediums are working, you have no way of maximizing the return on your advertising expenses.

MONITOR YOUR RESULTS

Keeping track of how your plan is working is critical to the success of your business. For example, assume you have placed four advertisements in four different publications. Before you ran the ads, business was slow. Now the phone is ringing and business is better. You know the ads increased your business, but do you know which ad worked best or which publication was responsible for the highest number of responses?

It is important to know why and how your actions generated new business. If you don't track the results of your advertising, you will waste a lot of money. When you ran the four ads, you should have keyed the ads. This simply means planting a device in the ad to identify the source of your new business. Mail-order businesses key their ads in many ways, but a favorite way is to add a letter to their box number.

For example, run your first ad with the box number 1029-A. The second ad might carry a box number of 1029-X. The third ad might be keyed as box number 1029-P. By keying the ads, you will be able to identify the source of the customer response. This lets you rule out publications that don't produce profitable results and to increase your advertising in the publications with the most pull.

If you run a service business where most work is requested by phone, keying your ads with a box number is not going to help you. You will have to come up with another way of identifying your customer source. You might try using a variation of your name. One ad might tell the reader to ask for John. The next ad could instruct the caller to ask for Mr. Woodson.

Offering a discount to the customer for mentioning the ad is another way to keep track of your advertising results. It is not uncommon to see an ad that offers a 10-percent discount to customers who mention an ad.

Putting a coupon in the ad is another way to find out the source of your business. Have your ad tell the reader to present the coupon at the time of service for a discount. This method is effective and gives clear evidence of where the customer got your name.

There is also the direct approach: ask the customer where they learned about your business. Most people will be glad to tell you where they heard of you and why they called you. This type of information is very valuable. It enables you to refine your marketing plan, increase your business, and save money that would be wasted in the wrong media.

USE ADVERTISING FOR MULTIPLE PURPOSES

Advertising is used primarily to generate consumer interest in goods and services. But advertising can do much more for a business. Advertising can build name recognition for your company. Advertising can be used to build your company image. Through advertising you can create almost any look you like for your business. If you want to be known as an expert in basement waterproofing, advertising can get the job done.

Build name recognition

You want people to see or hear your company name and feel like they know the company. To accomplish this goal, you must use repetitive advertising.

Take radio advertising for example. When you hear radio commercials, you normally hear the company name more than once. Pay attention the next time you hear ads on the radio. You will probably hear the company name or the name of the product being sold at least three times.

Television uses verbal and visual repetition to ingrain a name or product in the viewer's mind. Watch a few television commercials and you will see what I mean. During the commercials you will see or hear the company name or product several times.

Not only should your name be used often in the ad, the ad should run regularly. If you advertise in the newspaper, don't run one ad and stop. Run the same ad several different times. Use your logo in the ad, and publish the ads on a regular schedule. This type of repetition will implant your company name in the minds of potential customers. When these potential customers need your service, they will think of your company.

Generating direct sales

The need to generate direct sales activity is the reason most people use advertising. For a service business, generating direct sales is possible with direct mail, radio, television, print ads, telemarketing, and other forms of creative marketing. Telemarketing and direct mail are two of the fastest ways to generate sales activity.

We've already talked about how direct mail works, but how about telemarketing? Telemarketing is a job for thick-skinned people. Calling people you don't know and asking them to use your services, buy your product, or allow you into their home for a free inspection, estimate, or whatever is not much fun. However, if you can live with rejection and are not afraid to call 100 people to get 10 sales appointments, cold calling will work.

When your motive is to generate sales activity, it helps to make a time-restricted offer. Offer a discount for a limited time only. Create a situation where people must act now to benefit from your advertising. Turning up the heat with time-sensitive ads can generate activity quickly.

Sell to existing customers

Getting a customer can be an expensive proposition. If you were to calculate how much time and money you have invested for each customer, the results might shock you. Once you get these hard-earned customers, don't let them get away. First, take good care of your customers and keep them happy. Second, don't let them get away with just a single sale. Sell to your existing customers over and over. Use existing customers to generate leads on new customers.

Most customers that buy from you once will buy from you again. Send a mailing to each of your customers on a routine basis. Newsletters are often mailed each quarter to maintain customer contact and generate sales opportunities. Don't overlook the value of your customer base. Always keep your name fresh in the minds of your customers and never stop selling to them.

PLAN PROMOTIONAL ACTIVITIES

Promotional activities are an excellent way to get more sales and to build name recognition. Use special promotions to capture public attention and create an opportunity for additional sales. Let me give you an example of how you could stage a promotional event.

For this example, assume you are a contractor who specializes in kitchen and bath remodeling. Develop a seminar and ask the local material supplier to allow you to come into the store and give the remodeling seminar to shoppers. Tell the supplier how the seminar will be good for the store's image and increased material sales. Advertise the free seminar for about two weeks prior to the date of the event. The supplier may be willing to pay a portion of the ad costs because the store also is gaining publicity from this promotion.

When people begin gathering around you in the store, be sure to have your business cards, rate sheets, before-and-after photos, and other sales aids out where the shoppers can see them. After your seminar, field questions from the audience. This type of promotion can create the image of you being an expert in your field.

If it is legal in your area, give away a door prize. Have the audience fill out cards with their names, addresses, and phone numbers for a prize drawing. The prize could be a discount on remodeling services, a small appliance, or just about anything else. After the seminar, you will have a box full of names and addresses of potential customers who are already interested in your services and who know your name.

Advertising is a powerful business tool. In skilled hands, advertising can produce fantastic sales leads. Advertising can get you in the house, so you or your company representative can make the sale. Talk to some professionals in the field of marketing, and I think you will be surprised at the results you can achieve with advertising.

18 Employee issues

Finding the proper place for employees in your business may cause you to lose sleep at night. There is no question that employees will complicate your life. However, the right employees will make you more money than you could make alone. So, how will you decide on how to handle the employee issue? This chapter is going to open up the questions and show you the answers. You will learn the ropes of finding, hiring, managing, and terminating employees.

There are many factors that will influence your experiences with employees, but don't get the idea that more is always better. A lot of contractors do much better financially with small crews than they do with large crews. As we move through this chapter you will see specific examples of how employees can affect your business.

DO YOU WANT EMPLOYEES?

This is a simple question, and you shouldn't need any help to answer it. You either do or don't want employees. For now, we are not talking about your needs, we are discussing your desires. Assuming that you do want employees, let's examine a few justifiable questions.

Why do you want employees? Ah, now there's a question that requires a little thought. Do you want employees in order to make more money? Do you want employees to cast a brighter public image? Would you like to have employees so that you can feel more important? Will having employees make you more successful in your own mind? Well, let's look at each of these questions and evaluate your answers.

Why?

Many business owners can't provide a clear answer to this question. The owners may be quick to say they want employees, but they often can't express why they want them. Before you begin building a list of employees, decide why you are hiring them. Look at your reasons and be sure they make sense. It is easy to hire employees, but it is not as easy to get rid of them. Don't hire people until you need them and then know why you are hiring them.

To make more money?

Most employers will give this as their reason for hiring employees. There is a belief, true or false, that the more employees you have, the more money you will make. While this theory can hold true, it can also be dead wrong.

I have seen both sides. There have been times when I had several employees and made less in net profits than I did without them. Then there were occasions when employees were good for my financial health. What caused the differences? There were many factors in my personal experiences.

The economy has often affected the success of my business endeavors and my experience with employees. In my early years, poor management contributed to my failings with employees. I have owned various types of businesses and the type of business I was in at the time seemed to have an impact on the effectiveness of employees. My selection of employees has definitely made a difference in the success and failure of the relationship.

To cast a brighter public image?

It is true that the public often associates the success of a company with the number of employees the company has. But is this enough of a reason for you to hire employees? No. Hire employees for profitable reasons, not for public opinion. Yes, I know public image has an effect on the success of your business, but don't burden yourself with employees for only this one reason.

To make you feel important?

Most people won't admit to this, but many of them do feel more important when they have employees. This is not a good reason to hire people. There are less costly ways to improve your self appreciation.

To make you feel successful?

People measure success with different measuring sticks. Some people consider themselves successful when they have a lot of money. Having a number of employees can spell success for some individuals. Family health and happiness is a common measurement of success for business owners.

How you measure success is up to you, but don't lean too heavily on employees to find happiness. Simply having a large number of people on your payroll doesn't mean you will be happy. Unless you are satisfied providing jobs for others, while you get by on meager profits, employees can be a mistake.

I can see where you might be forming an opinion, but don't judge me too quickly. I'm not against employees, but I do want to show you both sides of the issue. Many business owners never take the time to consider the bad points of hiring employees. Since I assume you have dozens of reasons why hiring employees is great, I want to expose you to some other possibilities.

DO YOU NEED EMPLOYEES TO MEET YOUR GOALS?

The answer to this question may lie in your goals. There are limitations to what any one individual can do without help. However, you may be able to accomplish your goal with independent contractors and remove the need for regular employees. Independent contractors may seem to be more expensive than employees when their rates are first reviewed, but further investigation may prove the independent professionals to be less expensive. Let's look at the pros and cons of employees versus independent contractors.

Hourly rates

The hourly rates charged by independent contractors will normally be higher than the wages you would pay an employee to perform the same function. This is to be expected, but it may not be as it seems. While the hourly rate of independents is higher, they may actually cost less.

Since independents often work for multiple employers, they do not depend on you for their entire income. You pay only for what you need and for when you need it. Employees, on the other hand, rely on you to pay their salaries for the whole year. While you may be paying a lower hourly rate to employees, if you are paying them for time when you don't need them, you may be wasting money.

Insurance costs

Independent contractors pay for their own insurance. This can add up to a substantial amount of money. There is liability insurance, health insurance, dental insurance, worker's comp insurance, and disability insurance that you may have to pay for your employees, but you won't have these same expenses with subcontractors. You will still want liability insurance, but the rest of the coverages can be avoided.

Transportation

Subcontractors are responsible for their own transportation. If you hire employees, you will probably have to furnish company vehicles if they are field personnel. When you consider the cost of acquiring, insuring, and maintaining vehicles, the subcontractors who provide their own transportation are desirable.

Payroll

Payroll for companies with many employees can be a full-time job. If you have to pay a full-time employee just to handle the payroll for your other employees, you have additional overhead that will have to be recovered in the price of your services. By using subcontractors you eliminate the need for payroll and payroll records. Of course, you will still have to write checks to the subs, but this procedure is less labor intensive than doing payroll.

Payroll taxes

Payroll taxes are another expense you eliminate when you use subcontractors. When you have employees, you will have additional tax deposits to make for the payroll. This tax is nonexistent when independent contractors are utilized.

Paid vacations

Most employees expect to receive paid vacations. If you give the average tradesperson a 2-week paid vacation, you are losing over $1,000 a year. If you have 10 plumbers, you are losing over $10,000 a year by being a nice boss. Subcontractors don't expect you to give them a paid vacation. This type of savings can add up quickly.

Sick leave and other benefits

Sick leave and other benefits can be compared to paid vacations. Most employees expect these favors, but subcontractors don't.

Convenience

If convenience is a factor, employees may have an edge over subcontractors. Subcontractors can be difficult to control, after all why do you think they are called independent contractors. If you want people at your fingertips, employees are generally more reliable than subs.

Competition

Many contractors fear that their subcontractors will become direct competition. These contractors assume the independents are just using them to get to their customers. While some subcontractors will attempt to steal your clients, most won't.

While employees are not generally looked upon as competition, they may pose more of a threat than subcontractors. Subcontractors are already in business and already have customers. Employees may be looking to go into business for themselves. They also may be considering taking some of your customers with them. From the competition angle, either group is a possible threat, but I would be more concerned about employees.

Comparisons

Comparisons between employees and subcontractors are not hard to make. Before you dash out and hire employees, consider the advantages to using subcontractors. You may find that you are happier and more prosperous with the use of independent contractors.

WHERE DO YOU FIND GOOD EMPLOYEES?

Before you can hope to hire good employees, you must understand what makes a good employee. Defining the qualities you want in your staff should be easy. The problem will come when you try to find such an employee who is available for work. Most good employees are well cared for, and it is hard to pry them away from their present employers.

Classified ads

Classified ads are the quickest way to get applicants, but they generally are not the best way for you to find prime candidates for your opening. However, classified ads can produce top-notch people. If you go into your employee search with the knowledge you will have to sift through a mass of unqualified applicants to find one good worker, classified ads can be worthwhile.

Employment agencies

Employment agencies are known for their work with executives and professionals. These agencies are geared more toward white-collar positions than blue-collar jobs. Generally, the agency will contact you and go over the traits of a prospective employee before the employee knows who you are. Then, if you are interested in talking with the applicant, the agency will arrange an interview.

Some agencies charge applicants when they locate a job for them. However, most agencies charge employers for finding acceptable applicants. Before you deal with an agency, be sure of what you are getting. If you are asked to sign a contract or engagement letter, read it carefully and consider consulting your attorney before you sign the document. The fees charged by some agencies are ridiculously high.

Word-of-mouth

Word-of-mouth referrals are a good way to find the best employees. If you put the word out that you are looking for help, you may drum up some applicants through your existing employees or friends. When a person applies for a job because a friend has recommended your company, you have the advantage of built-in credibility.

Unemployment office

Unemployment offices carry listings of job opportunities for people in need of work. If you have a job to fill, notify the local unemployment office. They will put your opening into their computers and on their bulletin boards. This type of listing service is free and can produce quick results.

Handpicking

Handpicking employees is one of the best ways to get what you want. However, you must remain ethical in your procedures as you select and solicit individuals. Stealing employees from your competition is frowned upon. Let's look at how you should and shouldn't go after specific employees.

There is a right way and a wrong way to handpick employees. First, let's look at an example of the wrong way. Years ago I was a project superintendent on a townhouse project. I was in charge of all the plumbers and support people for the plumbers. During this time the economy was good and all good plumbers had jobs. However, most plumbing companies needed more plumbers to keep up with the rapid building trends.

There were companies that showed no remorse when they stole good plumbers from the competition. During this project I saw representatives from competitive companies come onto my job and offer my plumbers more money, right in front of me!

The plumbers knew they could get work anywhere, and many of them would change jobs for an extra 25¢ an hour. When these people raided my job, they often left with my plumbers in the backs of their trucks. The plumbers didn't give any notice; they just left, following the higher hourly wages. Some plumbers even tried to get me and the other company's representative into bidding wars.

You can imagine the ill will that formed between companies under these circumstances. One company would steal a plumber on Monday, and on Friday another company would take the plumber away to a new job. This endless turnover of plumbers hurt everyone in the business. You don't want to use these techniques when you handpick employees.

There are many ways to make employees aware that you are interested in offering them a position. One excellent way is to run into them at the supply house. While the two of you are standing around, waiting for your orders to be filled, start a conversation that leads to your need for help. Stress how you are looking for someone just like the person to whom you are talking. If the other person is interested in pursuing employment with your company, you should get some signals.

If you have your eye on a particular person for your position, don't hesitate to call or write the individual. If you don't want to look too obvious, ask the person if he knows of anyone with qualifications like his who is looking for work. By using this approach, you can give all the details you want about the position, without making a direct solicitation of the individual.

There are many tactful ways to get your point across to people that presently are employed. You don't have to stoop to going public with your attempt to take the employee away from the existing employer. Be discreet and keep your dealings fair.

EMPLOYEE POINTERS

There are specific laws that govern the employer-employee relationship. Let me give you a few pointers as a starting point. Don't take these pointers as the last word and don't consider them conclusive. Look upon them as a guide to the questions you should ask your attorney and tax professional.

Employment applications

Have all prospective employees complete an approved employment application. These applications tell you something about the person you are considering, and they provide a physical record for your files. Be sure the application forms you use are legal and don't ask prohibited questions.

W-2 forms

W-2 forms are used to notify employees of the total of their earnings during the past year and the amount of money withheld for various types of taxes. The forms must be mailed or given to employees no later than January 31 of the following year. A W-2 copy must be provided for the employee's federal and state returns as well as one for his records (FIG. 18-1). You must also file a W-3 Transmittal of Income and Tax Statements with federal and state tax authorities (FIG. 18-2).

W-4 forms

W-4 forms are government forms that must be completed and signed by each employee. These forms tell you, the employer, how much tax to withhold from the employee's paycheck. Once the form is filled out and signed, keep it in the employee's employment file. New W-4 forms should be completed and signed by employees each year.

I-9 forms

I-9 forms are employment-eligibility-verification forms. These forms must be completed by all employees hired after November 7, 1987. I-9 forms must be completed within the first three days of employment.

Employers are required to verify an employee's identity and employment eligibility. Employees must provide the employer with specific documents that substantiate eligibility. Some acceptable forms of identification

1 Control number	22222	For Official Use Only ▶ OMB No. 1545-0008	
2 Employer's name, address, and ZIP code		6 Statutory employee ☐ Deceased ☐ Pension plan ☐ Legal rep. ☐ 942 emp. ☐ Subtotal ☐ Deferred compensation ☐ Void ☐	
		7 Allocated tips	8 Advance EIC payment
		9 Federal income tax withheld	10 Wages, tips, other compensation
3 Employer's identification number	4 Employer's state I.D. number	11 Social security tax withheld	12 Social security wages
5 Employee's social security number		13 Social security tips	14 Medicare wages and tips
19a Employee's name (first, middle initial, last)		15 Medicare tax withheld	16 Nonqualified plans
19b Employee's address and ZIP code		17 See Instrs. for Form W-2	18 Other
20	21	22 Dependent care benefits	23 Benefits included in Box 10

24 State income tax	25 State wages, tips, etc.	26 Name of state	27 Local income tax	28 Local wages, tips, etc.	29 Name of locality

18-1 W-2 Wage and Tax Statement 1992. Department of the Treasury Internal Revenue Service

are birth certificates, drivers licenses, and U.S. passports. If an employer fails to comply with the I-9 requirements, stiff penalties may result.

1099 forms

1099 forms are used to report money you pay to subcontractors. These forms must be completed and mailed by January 31 of the following year. You will send one copy to the subcontractor, one copy to the tax authorities, and you will retain a copy for your files. Since subcontractors come and go, they can be difficult to locate when the time comes to send out the 1099 forms. Insist on having current addresses on all of your subcontractors at all times (FIG. 18-3).

Employee tax withholdings

Employee tax withholdings cannot be ignored. When you do payroll, you must withhold the proper taxes from each employee's check. The IRS will provide you with a guide that explains how to figure individual income tax withholdings. You will also have to deduct for Social Security (FICA) from each employee's paycheck. Figure this deduction using the tax guidebook available from the IRS.

Once you have computed the income and FICA withholdings, deduct them from the gross amount due the employee for wages (FIG. 18-4). After doing this, enter the amount of withholdings in your bookkeeping records.

DO NOT STAPLE

1 Control number	33333	For Official Use Only ▶ OMB No. 1545-0008	
Kind of Payer ▶	2 941/941E ☐ Military ☐ 943 ☐ CT-1 ☐ 942 ☐ Medicare govt. emp. ☐	3 Employer's state I.D. number	5 Total number of statements
		4	
6 Establishment number	7 Allocated tips	8 Advance EIC payments	
9 Federal income tax withheld	10 Wages, tips, and other compensation	11 Social security tax withheld	
12 Social security wages	13 Social security tips	14 Medicare wages and tips	
15 Medicare tax withheld	16 Nonqualified plans	17 Deferred compensation	
18 Employer's identification number		19 Other EIN used this year	
20 Employer's name		21 Dependent care benefits	
		23 Adjusted total social security wages and tips	
		24 Adjusted total Medicare wages and tips	
22 Employer's address and ZIP code (If available, place label over Boxes 18 and 20.)		25 Income tax withheld by third-party payer	

Under penalties of perjury, I declare that I have examined this return and accompanying documents, and, to the best of my knowledge and belief, they are true, correct, and complete.

Signature ▶ Title ▶ Date ▶

Telephone number ______________________

Form **W-3 Transmittal of Income and Tax Statements 1992** Department of the Treasury Internal Revenue Service

Please return this entire page with Copy A of Forms W-2 to the Social Security Administration address for your state as listed below. **Household employers filing Forms W-2 for household employees should send the forms to the Albuquerque Data Operations Center.**

Note: Extra postage may be necessary if the report you send contains more than a few pages or if the envelope is larger than letter size. Do NOT order forms from the addresses listed below. You may order forms by calling 1-800-829-3676.

If your legal residence, principal place of business, or office, or agency is located in	Use this address
Alaska, Arizona, California, Colorado, Hawaii, Idaho, Iowa, Minnesota, Missouri, Montana, Nebraska, Nevada, North Dakota, Oregon, South Dakota, Utah, Washington, Wisconsin, Wyoming	Social Security Administration Data Operations Center Salinas, CA 93911
Alabama, Arkansas, Florida, Georgia, Illinois, Kansas, Louisiana, Mississippi, New Mexico, Oklahoma, South Carolina, Tennessee, Texas	Social Security Administration Data Operations Center Albuquerque, NM 87180
Connecticut, Delaware, District of Columbia, Indiana, Kentucky, Maine, Maryland, Massachusetts, Michigan, New Hampshire, New Jersey, New York, North Carolina, Ohio, Pennsylvania, Rhode Island, Vermont, Virginia, West Virginia	Social Security Administration Data Operations Center Wilkes-Barre, PA 18769
If you have no legal residence or principal place of business in any state	Social Security Administration Data Operations Center Wilkes-Barre, PA 18769

Paperwork Reduction Act Notice.—We ask for the information on this form to carry out the Internal Revenue laws of the United States. You are required to give us the information. We need it to ensure that taxpayers are complying with these laws and to allow us to figure and collect the right amount of tax.

The time needed to complete and file this form will vary depending on individual circumstances. The estimated average time is 27 minutes. If you have comments concerning the accuracy of this time estimate or suggestions for making this form more simple, we would be happy to hear from you. You can write to both the **Internal Revenue Service**, Washington, DC 20224, Attention: IRS Reports Clearance Officer T:FP; and the **Office of Management and Budget**, Paperwork Reduction Project (1545-0008), Washington, DC 20503. Do NOT send the form to either of these offices. Instead, see the chart above for information on where to send this form.

Cat. No. 10159Y

18-2 W-3 Transmittal of Income and Tax Statements 1992.

Department of the Treasury—Internal Revenue Service

9595 ☐ VOID ☐ CORRECTED

PAYER'S name, street address, city, state, and ZIP code		1 Rents $	OMB No. 1545-0115	Miscellaneous Income
		2 Royalties $	1992	
		3 Prizes, awards, etc. $		
PAYER'S Federal identification number	RECIPIENT'S identification number	4 Federal income tax withheld $	5 Fishing boat proceeds $	Copy A For Internal Revenue Service Center
RECIPIENT'S name		6 Medical and health care payments $	7 Nonemployee compensation $	File with Form 1096.
Street address (including apt. no.)		8 Substitute payments in lieu of dividends or interest $	9 Payer made direct sales of $5,000 or more of consumer products to a buyer (recipient) for resale ▶ ☐	For Paperwork Reduction Act Notice and instructions for completing this form, see Instructions for Forms 1099, 1098, 5498, and W-2G.
City, state, and ZIP code		10 Crop insurance proceeds $	11 State income tax withheld $	
Account number (optional)	2nd TIN Not. ☐	12 State/Payer's state number		

Form **1099-MISC** Cat. No. 14425J Department of the Treasury - Internal Revenue Service

18-3 Form 1099—Miscellaneous Income. Department of the Treasury—Internal Revenue Service

While the money withheld is still in your bank account, it doesn't belong to you. Don't spend it!

As an employer, you must contribute to the FICA fund for each employee. Your contribution will be equal to the amount withheld from the employee's paycheck. This is just another hidden expense associated with having employees. Your portion of the funding will do you no good, it only benefits the employee.

Employer ID number

When you establish your business, you should apply to the IRS for an employer ID number. Some small business owners use their social security numbers as an employer ID number. It is better to receive a formal ID number from the IRS. You will use this identification number when you make payroll-tax deposits.

Payroll-tax deposits

Payroll-tax deposits are the money you withhold from your employee's paycheck and your matching FICA payment. Payroll-tax deposits are required of all companies with employees. The IRS will provide you with a book of deposit coupons. The deposits can be made at your bank. The requirements for when these deposits must be made fluctuate from business to business. Consult your CPA for precise instructions on when you must make your payroll-tax deposits.

As a business owner, you can be held personally responsible for unpaid payroll taxes. Even if you sell or close your business before payroll taxes are paid, the tax authorities can come after your personal assets to settle the debt. Don't play around with the money owed on payroll taxes.

Social Security Benefits Worksheet—Lines 21a and 21b (keep for your records)

If you are married filing a separate return and you did **not** live with your spouse at any time in 1991, enter "D" on the dotted line next to line 21a.

1. Enter the total amount from **Box 5** of **all** your **Forms SSA-1099** and **Forms RRB-1099** (if applicable) . . . . 1. ______

Note: *If line 1 is zero or less, stop here; none of your benefits are taxable. Otherwise, go to line 2.*

2. Divide line 1 above by 2 . . . . 2. ______
3. Add the amounts on Form 1040, lines 7, 8a, 8b, 9 through 15, 16b, 17b, 18 through 20, and line 22. Do not include here any amounts from Box 5 of Forms SSA-1099 or RRB-1099 . 3. ______
4. Add lines 2 and 3 . . . . 4. ______
5. Enter the total adjustments from Form 1040, line 30 . . 5. ______
6. Subtract line 5 from line 4. . . . . 6. ______
7. Enter on line 7 the amount shown below for your filing status: 7. ______
   - Single, Head of household, or Qualifying widow(er) with dependent child, enter $25,000
   - Married filing a joint return, enter $32,000
   - Married filing a separate return, enter -0- ($25,000 if you did **not** live with your spouse at any time in 1991)
8. Subtract line 7 from line 6. Enter the result but not less than zero. 8. ______
   - If line 8 is zero, stop here. None of your benefits are taxable. Do not enter any amounts on lines 21a or 21b. **But** if you are married filing a separate return and you did **not** live with your spouse at any time in 1991, enter -0- on line 21b. Be sure you entered "D" on the dotted line next to line 21a.
   - If line 8 is more than zero, go to line 9.
9. Divide line 8 above by 2 . . . . 9. ______
10. **Taxable social security benefits.**
   - First, enter on Form 1040, line 21a, the amount from line 1 above.
   - Then, enter the **smaller** of line 2 or line 9 here and on Form 1040, line 21b. . . . . 10. ______

Note: *If part of your benefits are taxable for 1991* **and** *they include benefits paid in 1991 that were for an earlier year, you may be able to reduce the taxable amount shown on the worksheet. Get Pub. 915 for details.*

18-4 Social Security Benefits Worksheet. Department of the Treasury—Internal Revenue Service

Federal Unemployment Tax

The Federal Unemployment Tax is also known as FUTA. Your requirements for making FUTA deposits will depend on the gross amount of wages paid to your employees in a given period of time. Talk to your accountant for full details on how FUTA will affect your business.

Self-employment tax

As your own boss, you will have to pay self-employment taxes. Your FICA tax rate may be nearly double. This is done to make up for the fact that you don't have an outside employer contributing to your portion of the Social Security fund.

State taxes

Different states have different tax requirements. To be safe, check with your CPA or local tax authority to establish your requirements under local tax laws.

Labor laws

The labor laws control such areas as minimum-wage payments, overtime wages, child labor, and similar issues. As an employer you must adhere to the rulings set forth in these laws. The government will be happy to provide you with information on your responsibilities. Call the Department of Labor to request information and materials.

OSHA regulations

The Occupational Safety and Health Act (OSHA) controls safety in the workplace. Your business may be affected by OSHA in many ways. To learn the requirements of OSHA, contact the Department of Labor.

Terminating employees

With so many employee rights, you must be careful when you are forced to fire an employee. Since you never know when termination will be your only option in dealing with a troublesome employee, you should assume all employees are possible targets for termination. By this I mean you should create and maintain a paper trail on each employee's activities.

As soon as you hire an employee, start an employment file on the individual (FIGS. 18-5 and 18-6). The file will grow to contain all documentation you have on the employee. The file should include: tax forms, employment application, income records, performance reviews, attendance records, and disciplinary actions and warnings. If a time comes when you must dismiss an employee, these records of employment history may come in handy.

Before you lose your temper and fire an employee, consider the costs you will incur replacing the worker. Give yourself time to think about the offense. Is it really necessary to fire the individual? If the circumstances demand termination, do so with care. Consult with your attorney in advance to be certain of your requirements before letting an employee go.

HOW TO KEEP GOOD EMPLOYEES

If employees are worth having, someone else will want them. You will always run the risk that someone will try to persuade your best employees to leave your company for theirs. You should also be concerned about good

Employee-File Checklist

Employee name: ______________________________

Employee ss # ______________________________

Item	*In file*	*Need*	*Notes*
I-9 Form			
W-4 Form			
Application			
Tax Info			
Insurance Info			
Reviews			
Warnings			
Attendance			

18-5 Employee-file checklist.

Weekly Work History

Employee: ______________________________

Payroll number: ______________________________

Date	*Work phase & Job name*	*Time in*	*Time out*	*Total time*

18-6 Weekly work history form.

employees going into business for themselves, and into competition against you. You can't really blame the employees for wanting their own business, after all, you wanted your own business.

What you have to do is make the working conditions so good that the employees won't want to leave. Many factors will influence employees to stick with your company. Some of these factors are:

- comfortable wages
- health insurance
- dental insurance
- paid vacations
- sick leave
- company vehicles
- retirement plans
- good working conditions
- a friendly atmosphere
- competent coworkers
- fair supervisors
- pride in the company

If you establish a good environment for your employees, they will have no reason to leave your employ. Bonus plans and other incentives can even remove much of the risk of having the employees going into business for themselves. It will be up to you to communicate with your employees and to create circumstances that will keep them happy. If you have valuable employees, they are worth the extra effort.

CONTROL EMPLOYEE THEFT

All business owners would like to believe that their employees aren't thieves, but some are, and you have to protect yourself and your good employees from the few bad ones. Most employee theft is petty, but that doesn't mean it is not serious. Stealing is stealing, and you can't afford to have criminals for employees.

How you run your business will have a bearing on the amount of employee theft. If you thoroughly screen all of your job applicants, you can cut down on the chances of hiring a crook. By keeping tight control on your inventory and making your employees aware of your antitheft policies, you can reduce your risks even more. If you eliminate temptation, you eliminate most casual theft.

EXERCISE QUALITY CONTROL

Quality control is just what it sounds like—controlling quality. The qualities you control could be numerous. The qualities most business owners are interested in controlling are:

- punctuality
- good work habits
- customer service
- work quality
- dependability
- loyalty

There are, of course, other qualities you may wish to keep in check. You might request that your employees take ongoing continuing education programs. Having your employees expand their capabilities into other work areas could be one of your pet projects. Once you know what you want from your employees, work with your employees to meet your goals.

TRAIN EMPLOYEES TO DO THE JOB

Today, most employers are looking for experienced people that can step into a position and be productive. The days of training apprentices is all but gone. One reason for this is money; it costs money to train employees. Even if you do the training yourself, it costs you money. The time that you spend away from your routine duties is lost income. To train employees, you must look upon the training as an investment.

In the old days, employees stuck with their employers for a long time. If an employer trained an employee, he could be reasonably confident that his investment would pay off. Today employees change jobs without blinking an eye. Employers know this and are reluctant to train employees that will run to another employer if the opportunity arises. It is sad, but the traditional values that once existed have been eroded with the increased demand for the mighty dollar.

The quest for money is not only in the minds of employees. Many employers don't want to hire inexperienced help because they know the new employees will not make as much money for them as an experienced employee. These circumstances are changing the business world. Since the old masters are not passing their knowledge down to apprentices, the crop of qualified tradespeople is shrinking.

I guess it makes more sense to hire people that can jump out there and start turning a dollar. But, if you do train employees to produce the type of work you want, you may be happier. There is a certain satisfaction gained from watching a rookie mature into a journeyman. The choice is yours, but be advised, trainees are likely to look for a higher-paying job once you have trained them.

TRAIN EMPLOYEES TO DEAL WITH CUSTOMERS

Every business is run differently, and even experienced mechanics will have to be taught to treat customers the way you want them treated. You may choose not to provide on-the-job training for work skills, but don't forego training your employees to deal with customers. Your customers are your business. If you alienate them, you lose your business. Employees are representatives of your company. If they act improperly around customers, it will be a reflection on your business.

It is a good idea to develop a policy manual on how you want customers treated. Issue the manual to each of your employees and require them to commit it to memory. If necessary, test the employees' knowledge of the manual. Before you put people in touch with your customers, make sure they will behave in a suitable manner.

ESTABLISH THE COST OF EACH EMPLOYEE

There are many hidden costs involved with employees, and each employee may have a different set of circumstances. Before you set your prices, know what each of your employees is costing you.

The most obvious cost is the hourly wage, but there are other factors. Some employees will receive more benefits than others. If one employee gets a two-week paid vacation and another employee gets one week of paid vacation, the cost of the employees will be different by an amount equal to the extra week of vacation pay.

As employees build seniority, they normally gain additional benefits. You must consider all of these costs when you determine the overall cost of an employee. Whether it is health insurance, dental insurance, or paid leave, you must factor the cost into your projections.

Bonus pay is another item that can influence the cost of your employees. If you are in the habit of giving each employee a bonus during the holiday season, you must count this money in the cost of the employee.

Turn over every rock to look for hidden expenses. Don't let any part of the expenses for your employees go unnoticed. Once you have all the figures, chart the hourly differences. When you bid a job, bid it based on your most expensive employees. Then if you can put employees that cost less on the job, you make more money. But if your least expensive labor is not available, you will not lose anything by putting your top-paid people on the job.

DEAL WITH DOWNTIME

You have already seen how you can lose money if your crews must stop to run for materials, but that is not the only way you can lose money to downtime. Some causes for these losses will be beyond your control, but many of them can be avoided with strong management skills.

Bad weather

Bad weather can often shut down a contractor. While you can't control the weather, you can plan for its effect on your business. If you have a business that involves some inside work, try to save this work for days when the weather won't allow your normal outside operations. If you are starting a job where the weather might cause delays, plan on ways to circumvent the lost time. This might involve using tarps to cover the work area or renting heaters to keep the job comfortable. Look for ways to keep production up during any weather conditions.

When the circumstances cannot be overcome, use your best judgment. Most employers will send their employees home without pay. On the surface this saves money, but it may cause you to lose your employees. Good employees are hard to find, and a turnover in employees is expensive. You may be money ahead to create some busywork for the crews, even if it is not cost effective.

Some ideas for busywork might be counting inventory, taking trucks in for service, or performing maintenance on equipment. While these tasks may not warrant the use of highly-paid personnel, they must be done, and if it keeps your employees in place, you may be better off.

Past-due deliveries

Past-due deliveries can bring your crews to a halt. If you are a good manager, you won't let this happen. However, there will be times when a delivery isn't made, and you must find work for your crews. Be prepared for these times with some backup plans. If you send the crews home, they may not be happy. On the other hand, maybe they would enjoy having the day off, even if they aren't getting paid. Give them the option of taking the day off or doing fill-in work.

Code-enforcement rejections

Code-enforcement rejections can bring work to a sudden stop. It is difficult to think of a suitable excuse for this type of downtime. If you or your field supervisors are supervising the work, you should not fail an inspection. If you have recurrent problems of this nature, you need tighter field supervision.

Disabled vehicles

Every contractor is going to have problems with disabled vehicles from time to time. The most you can do to prevent these problems is regular maintenance. When a truck breaks down and is going to be out of service for an extended time, try to double up your crews. There isn't much else you can do.

Work lulls

Sooner or later, a lull in work will cause you downtime with your crews. There are times of the year when these lulls can be projected: holidays, summer vacation seasons, tax seasons, and school start-up seasons. Proper preparation can help you overcome these slow periods. Line up work in advance for the slow times. Aggressively advertise and offer discounts, if necessary, to keep your people busy. Avoid laying your people off. Once they are gone, you may not get them back.

REDUCE CALLBACKS AND WARRANTY WORK

Customers will not pay you to do the same work twice, but you will have to pay your employees for their time. This can get expensive fast. If you have sloppy workers that frequently cause callbacks, you must take action.

Everyone is going to make mistakes, but professionals shouldn't make many. Callbacks are generally the result of negligence. Either the mechanic did the job too quickly, too poorly, or didn't check the work before leaving the job. You can and must control this type of behavior.

Callbacks and warranty work hurt your business in two big ways. The first hurt is financial; you lose money on this type of work. The second problem is the confidence your customers lose in the quality of your work. You cannot afford either of these results. There are several options available for controlling these costly occurrences. Let's look at some of the ways that have worked for others and that may work for you.

Callback boards

Callback boards can reduce your callbacks if you have multiple employees. Hang a callback board in a part of your office that all employees can see. When a mechanic has a callback, the mechanic's name is put on the board. The board is cleared each month, but people with callbacks must see their name on the board for up to a full month.

Generally there is a certain competitiveness among tradespeople. If a mechanic's name is on the callback board, he will probably be embarrassed. This simple tactic can have a profound effect on your callback ratio.

Employee participation

Employee participation in the financial losses of callbacks is another option. However, your employees must agree to this plan without being pressured. For your protection, have all employees agree to the policy in writing.

Under the employee-participation program, employees agree to handle their callbacks on their own time. You pay for materials and the employees absorb the cost of the labor. As a variation of this program, you can agree to pay the employee for the first two callbacks in a given month, with the employee taking any additional callbacks without pay. But, before you implement either of these programs, confirm their legality in your area and have employees agree to your employment terms in writing.

Bonus incentives

Bonus incentives are another way to curtail callbacks. Callbacks are expensive and detrimental to your business image. If you can eliminate warranty work by offering bonuses, do it. You won't lose anymore money and you won't lose any credibility with your customers.

If you don't like the idea of giving employees bonuses for doing a job the way they should in the first place, hedge your bets. Determine what the maximum annual bonus for any employee will be and adjust your starting wages to build in a buffer for the bonuses. The employee will feel rewarded with the bonus and you won't be paying extra for services you expect to get out of a fair day's work.

MANAGING OFFICE EMPLOYEES

Office employees are a little easier to manage than field employees. People who work in your office are easier to find and to watch. If you are an office-based owner, your office employees will feel compelled to stay busy;

they know you are watching their performance. However, you must not abuse the power your presence presents.

Office employees can be intimidated by having the boss close at hand. If this happens, production will drop off or mistakes will multiply. You should hire the best help you can find and then let them do their jobs. If you are constantly looking over their shoulders, you will be doing more harm than good.

A common mistake made by first-time bosses is their involvement with office help. If you get bored, don't start bending the ear of your office help. When you distract the office workers, your work is not getting done. You must set an example for your employees. If they see you hanging around the coffeepot swapping stories, they will feel cheated that they don't have the same privileges. If you want to goof off, do it behind closed doors.

When you establish your office employees, don't neglect their needs and desires. If you have a good employee that wants a new chair, buy a new chair. When your workers want a coffeemaker, buy a coffeemaker. If your help's requests are reasonable, attend to them. Happy employees are more productive, not to mention nicer to be around.

MANAGING FIELD EMPLOYEES

Field employees present more management challenges than office help. These employees are mobile and can be difficult to monitor. You may know about every trip your secretary makes to the snack area, but you will be hard pressed to keep up with how many times your field crews take a break.

By monitoring job production, you can keep tabs on your crews. If the work is getting done on time, what difference does it make if the crew took three breaks, instead of two? If you have good employees that are turning out strong production, leave them alone. If you are concerned about your field crews, talk to your customers. Customers are generally very aware of how crews act. Make some unannounced visits to the job sites. Don't let the crews get too comfortable, but don't crowd them either.

SHOULD YOU ENLIST COMMISSIONED SALESPEOPLE?

This is a good question. The answer is determined by your business goals. Commissioned salespeople can make a dramatic difference in your business. On the plus side of the deal, commissioned salespeople can generate a high volume of gross sales. Since you are paying the sales staff only for what they sell, an army of sales associates can be mighty enticing. However, a high volume of sales can create numerous problems. You may not have enough help to get the jobs done on time. You might have to buy new trucks and equipment. The increased business may tie you to the office and cause your field supervision to suffer. There are many angles to consider before bringing a high-powered sales staff online.

The right salespeople might increase your business. With commissioned salespeople, you don't have the normal employee overhead and you only pay for what you get. Good salespeople will hustle up deals that would otherwise never come your way. A strong closer will make deals happen on the spot, so you have quick sales and no downtime. Sales professionals can take a simple estimate and turn it into a major job for your tradespeople. With the right training and experience, sales professionals can get more money for a job than the average contractor. It is clear that for some businesses a sales staff is a powerful advantage.

The drawbacks to commissioned salespeople may outweigh the advantages. Some salespeople will tell the customer anything they want to hear to get a signature on the contract. As the business owner, you will have to deal with this form of sales embellishment at some point. The customer might tell you that the salesperson assured him he would get screens with his replacement windows, when you had not figured screens into the cost of the job. The salesperson might have promised that the job could be done in two weeks, when in reality, the job will take four weeks. This type of sales hype can cause some serious problems for you and your workers.

Most sales associates are not tradespeople. They don't know all the ins and outs of a job. They know how to sell the job, not how to do it. A salesperson might tell a prospect that putting a bathroom in the basement is no problem, when in fact such an installation might require a sewer pump that adds nearly $800 to the cost of an average basement bath. There are many times when an outside sales staff undersells a job. Sometimes they sell the job cheap to get a sale. Other times the wrong price may be quoted out of ignorance. In either case, you as the business owner will have to answer to the customer.

Getting too many sales too quickly can be as devastating as not having enough sales. If the salesperson you put in the field is good, you might be swamped with work. This can lead to problems in scheduling work, the quality of the work turned out, field supervision, cash flow, and a host of other potential business killers.

Sending the wrong person out to represent your company can have a detrimental effect on your company image. If the salesperson is dishonest or gives the customer a hard time, your business reputation will suffer.

If you decide to use commissioned salespeople, I suggest you go with them on the first few sales calls. When you are interviewing people to represent your company, remember they are sales professionals. These people will be selling you during the interview with the same tenacity they will use on prospects in the field. Go into the relationship with your eyes wide open. Don't take anything for granted and check the individuals out for integrity and professionalism.

EMPLOYEE MOTIVATION TACTICS

There are many ways to motivate your employees. There are books written for that express purpose. A creative employer can always find ways to encourage employees to do better. Let me give you just a few suggestions that might work for your employees.

Awards

Awards are welcomed by everyone. You can issue award certificates for everything from perfect attendance to outstanding achievements. These inexpensive pieces of paper can make a world of difference to employees. An employee that knows he will get a certificate for coming to work every day will think twice before calling in sick when he isn't. While the award may not have a financial value, it becomes a goal. Employees that are working towards a goal will work better.

Money

Money is a great motivator. Since most people work for money, it stands to reason they may work a little harder for extra pay. Any type of bonus program benefits the production rate of your employees.

Day off with pay

Sometimes a day off with pay is worth much more to an employee than the value of the wages. This special treat could become a coveted goal. Hold a contest where the most productive employee of the month gets a day off with pay. Sure, you'll lose the cost of a day's pay, but how much will you gain from all of your employees during the competition?

Performance ratings

Performance ratings can be compared to awards. If employees know they will be rated on their performance, they may work harder. These ratings should be put in writing and kept in the employees' files.

Titles

Wise business owners know that a lot of people would rather have a fancy title than extra money. In fact, many companies promote people with new titles to avoid giving higher income raises. Even if your company is small, you can hand out some impressive titles. For example, instead of calling your field supervisor a foreman, call him a field coordinator. Instead of having a secretary, have an office manager. When you have someone that enters data into a computer all day, change the title from data entry clerk to computer operations executive. Titles make employees feel better about themselves, and they don't cost you anything.

19

Insurance, benefits, and retirement plans

Planning and purchasing insurance, benefits, and retirement plans can be very perplexing. These areas of your business are not simple, and the responsibilities for you as a business owner are imposing. As a business owner, when you think of insurance, you must consider all aspects of the issue.

If your mind is on health insurance, you must acknowledge the fact that you no longer have deductions taken from your paycheck and coverage provided. You must establish your own insurance program, pay all the costs, consider tax consequences, and determine what impact employees will have on the program you choose.

If you have employees or plan to hire employees, benefit packages are a serious consideration. If you don't offer employee benefits, you may not get or keep the best employees. It has become standard practice for employers to provide their workers with benefits.

The task of establishing and administering benefit packages can get complicated. There are many options available, each with its own advantages and disadvantages. If you are not knowledgeable of how the laws and rules regulate your actions as they pertain to benefits, you can get in a lot of trouble, fast.

Whether you are looking at retirement plans for yourself or your employees, the possibilities can be mind-boggling. Setting up a plan for yourself is one thing, establishing programs for employees is another.

COMPANY-PROVIDED INSURANCE FOR YOU

Putting an insurance program in place for yourself, when no other employees are involved, is not difficult. However, choosing the right plans will take some research. Let's take a closer look at each type of coverage and see how they fit into your business plans.

Health insurance

Health insurance is expensive and the plans are complex. Deciding on the best type of insurance will require research and thought. What should you look for in health insurance programs? Let's find out.

Pre-existing conditions Pre-existing conditions can have an influence on your choice of health insurance. Most insurance companies will not cover expenses related to a pre-existing condition. For example, if you have problems with your back when you obtain your new insurance, the insurance company may refuse to cover medical expenses incurred for back problems. If you have had a pregnancy that involved surgery or medical attention beyond the normal childbirth requirements, a reoccurrence of these circumstances may not be covered by your new policy. It is possible to obtain insurance coverage that does not eliminate pre-existing conditions. The price for these policies may be higher, but the protection may be worth the additional cost.

Deductible payments The deductible payments for insurance plans vary. Typically, the more you pay in deductible expenses, the lower the monthly premiums. A plan with a $200 deductible will cost more on a monthly basis than a plan with a $500 deductible. It is generally considered wise to choose a plan with a higher deductible and lower installment payments.

Limits of standard coverage Before you buy any insurance plan, understand the limits of standard coverage. Not all policies cover all possible circumstances. Read policies closely and ask questions. The insurance company may not have to disclose facts to you unless they are asked as direct questions.

Waiting period It is possible that an insurance policy will require a waiting period. These waiting periods stipulate that a specific amount of time must pass before a procedure is covered. The waiting period eliminates the risk to the insurance company. Determine if the policy you are considering has a waiting period and if so, what conditions apply to the rules of the waiting period.

Copayments Average health plans call for the insured to make copayments. This means that you will be responsible for paying a portion of your own medical expenses, even though you are insured. A common copayment amount is 20 percent of the costs incurred. You pay 20 percent and the insurance company pays 80 percent. However, the split can vary. You might find that you are responsible for 30 percent of the bills.

Verify how your intended policy deals with copayments. Some types of coverage are much more generous and pay nearly the entire cost of your medical expenses. For example, you may only pay a few dollars for each office visit to your doctor. These pay-all policies cost more, but they provide excellent coverage and you will not have to come up with large sums of out-of-pocket cash unexpectedly.

Dependent coverage If you have dependents, you will be interested to know how a policy deals with dependent coverage. Will your dependents receive the same coverage as you? Will the premiums be set at reduced rates for the additional coverage? Are there limits on dependent coverage? Is there an age limit on the coverage extended to your dependents? All of these are questions you should ask about dependent coverage.

Rate increases Rate increases are a fact of life with insurance. However, some insurance policies are more prone to rate increases than others. Ask how often the insurance company is allowed to raise its rates. Will you be faced with increases quarterly, semiannually, or annually? Inquire about caps on the amount of increase at any one interval. For example, if your rates will be subject to an increase on an annual basis, how much is the maximum the rate can be elevated? With insurance, you can never ask too many questions.

Group advantages As a business owner you may be eligible for group advantages. Some insurance companies will take small groups of customers and create a large group. This type of grouping is designed to offer coverage at lower rates. Normally your company will need at least two employees for this type of coverage. The savings may be worth putting your spouse on the payroll. Check with your insurance representative for the requirements for joining a group plan.

Dental insurance

Dental insurance is a blessing for people with bad teeth. If you have paid for crowns or root canals lately, you know they aren't cheap. This type of insurance is shunned by some and coveted by others. The decision is yours, but dental insurance can be well worth its cost for the right people.

When you shop for dental insurance, ask the same questions you ask for health insurance. Like health insurance, dental insurance comes in many forms. Choosing the right policy will be a matter of your personal needs.

If you decide to buy dental insurance, expect to go through a waiting period for major-expense coverage. While some policies will pick up immediately routine maintenance of your teeth, you will probably have to wait for those needed crowns and caps. The waiting period for major work is usually one year.

Once you are covered and eligible for payments on major work, don't expect the coverage to pick up the whole tab. Many dental plans will pay no more than one half of your major expenses. For example, if you are getting a $400 crown, your insurance might pay only $200.

Disability insurance

Disability insurance provides protection against lost income due to injuries and sudden disabilities. Short-term disability policies are designed to provide assistance for a short period of time. Long-term disability will continue to make payments for an extended period of time.

Disability policies provide you with a percentage of your normal income while you are unable to work. The percentage of your income that is paid will depend on your policy. These policies may also have a pre-existing condition waiver. Let me give you examples of how each type of disability plan might work.

Short-term disability Short-term disability polices will set a limit on the amount of time you can receive benefits. Six months is a common benchmark for the maximum period of time you may collect from a short-term policy.

There is usually a short waiting period before the disability income (DI) checks start. In most cases you will have to be out of work for at least a week before you can collect on your DI. The amount you can collect will be a percentage of your normal income. A plan that pays up to 50 percent of your income is not unusual. However, there is generally a maximum dollar amount that you can collect.

For example, your policy may pay 50 percent of your normal weekly pay, but it might stipulate that the maximum you can receive in any given week is $150. Obviously, if you make more than $300 a week, and most contractors do, you will not be getting half of your income in benefits. Watch out for these little stingers.

Long-term disability Long-term disability works on a similar principle as short-term disability. These plans may pay a higher percentage of your income than short-term DI. There will be limits on the minimum and the maximum monthly payments, but the length of time you can collect payments is frequently unlimited.

Life insurance

Life insurance doesn't seem very important until you have dependents. While you are single there is no need to worry about how people will get along without your income when you die. Your mind isn't filled with questions about how your bills will be paid after you are gone or how your child will grow up and be educated. However, when people you care for will be left behind after your death, life insurance becomes important.

How much life insurance do you need? What type of life insurance will suit your needs best? These are the two most commonly asked questions about life insurance. Every person might have a different correct answer. Life insurance must be tailored to your personal requirements.

The amount of life insurance coverage you need will depend on several factors. The first factor is the number of dependents you will leave behind. A person with only a spouse will need less insurance than a person with a spouse and two children. Another factor is your income. Many people suggest buying insurance coverage based on a multiple of your annual income. Some people say insurance benefits equal to your annual salary is enough. More people are inclined to believe it is better to have coverage equal to three year's worth of income. If you are leaving behind a spouse and children, the spouse may not have the earning ability to support the remaining family members. If this possibility exists, you should carry enough insurance to allow for investments and long-term support.

If your spouse isn't working and hasn't worked for some time, she may find it difficult to get a job. If you have been the sole provider, your spouse will have to grieve, adjust to your death, find work, and establish a new life. This is not only stressful, it takes time. Can all of this take place in one year? It could, but it would be a strain. So if you leave behind only one year's worth of benefits, your spouse will be under extra pressure. And don't forget, there will be burial costs, personal debts, and business expenses to be paid out of the benefits you bequeath. This consideration must be weighed when you determine the amount of your life insurance.

As you can see, there are numerous factors that you must consider when you determine the face amount of your life insurance. Some people look at life insurance as a one-time shot in the arm for their distressed family members. These people assume leaving their spouse $100,000, in cash, is more than adequate. In this mind-set, the spouse is expected to live off the $100,000 until a new life is built. This isn't a bad plan, but there is another perspective to consider.

In my estate planning, I have structured a way for my wife and daughter to derive most of their annual income needs from the interest of my life insurance dividends. When I die, if the proceeds from my life insurance are wisely invested, the passive income generated will be substantial. This passive income will support my family, without them having to deplete the lump sum of the premium payoff.

With this plan, my wife and daughter are well cared for, and the money paid by the insurance company remains virtually untouched. When my wife passes on, her life insurance dividends can be handled in a similar way. The end result for our daughter will be a comfortable income from her investments and a sizable nest egg in cash.

Of course, to generate this type of insurance payoff you have to carry some steep premiums. Not all people are willing to invest their money in life insurance, and I'm not saying you should. I believe you should buy as much life insurance as you feel you need, and not a penny more.

Term-life insurance Term insurance is fine as a supplemental life insurance, but it may not be the best choice as a primary insurance. When you are in your prime earning years and building assets, term policies can protect your family from incurring your debts. If you depend on term life insurance as your only life insurance, you may be distressed in later years. As you grow older, your premiums will go up and the value of the term policy will go down. If you live a normal life, the policy may not be worth much at the time of your death.

Whole-life policies Whole-life policies are more expensive than term insurance, but they are more dependable. The face amount of these policies doesn't decrease and the premiums don't go up.

There are other advantages to a whole-life policy. As you make monthly payments you build a cash value in the policy. In effect, you are creating a savings account of sorts. Later in life, if you need some quick cash you can borrow against your built-up cash value. Interest rates on these loans are usually very low, and you can pay back the money at your discretion.

If you reach a point in life where you no longer want to maintain your life insurance, you can cash in a whole-life policy and receive the cash value. These policies are considered one of the best available for the long haul.

Universal and variable policies Universal and variable life policies are variations of whole-life policies. These policies feature investment angles for your premium dollars. As you pay premiums, you build cash value and your account earns interest. The interest you earn is rolled over and is not taxed, unless it is withdrawn. Many business owners choose these policies.

Key-man insurance If you have been in business, you have probably heard of key-man insurance. This is a form of life insurance that protects a company against the death of a vital employee. Normally the employee is insured by the employer and the employer pays the insurance premiums. If the employee dies, the proceeds of the insurance goes to the employing company. This allows the company to have a cash buffer until the key employee can be replaced.

Unless you are in a partnership or a corporation with other stockholders, you shouldn't need key-man insurance. Regular life insurance can protect your family and cover your business debts. However, if you have a partner that you depend upon heavily, you might want to set up a key-man plan.

Other options for life insurance There are many other life insurance options. The abundance of plans is almost overwhelming. There are all types of riders that can be added to standard policies, and terms and conditions can be adjusted to meet every conceivable need. Due to the complexity of insurance programs, you should talk to several insurance professionals before you make a decision.

CHOOSING AN INSURANCE COMPANY

Choosing an insurance company is no easy job, but it might be the most critical aspect of your insurance planning. No one wants to pay insurance premiums for years only to have the insurance company go out of business. Not all insurance companies have the same financial strength. The investment abilities of some companies are much better than those of others.

Choose your insurance firm carefully. Research the company and attempt to establish its financial power and track record. By talking with your state agencies and going to major libraries, you should be able to find performance ratings on the various companies. Dig deep into a company's background before you depend on them to protect you.

PACKAGING EMPLOYEE BENEFITS

Employee benefits can be even harder to decipher than your own benefits. The rules and regulations that go with providing benefits to your employees make the chore challenging. As an employer you are responsible for the compliance of some rather strict laws and regulations. If you fail to execute your duties in the proper manner, you can wind up in serious trouble.

Many employers choose an insurance company that offers multiple benefits in a single plan. The benefits can include coverage for: medical, dental, life, disability, and accident insurance. This type of employee package might seem cost prohibitive, but it is an attractive feature when you are trying to hire and keep top-notch employees.

Some of these multi-plans allow employees to choose some of the types of coverage they want. The employer gives each employee a set allowance to allocate to various types of coverage, then the employee is free to customize his or her individual plan. This type of employee package is often referred to as a flexible benefit package and is sometimes called a cafeteria plan.

Other benefits you might offer your employees include: paid sick leave, paid personal days, paid vacation, retirement plans, and bonus programs. Retirement plans for you and your employees will be discussed in detail later in the chapter.

Before you decide to give benefits to your employees, research the rules and regulations you must follow. Talk to your attorney, your insurance agent, and your state agencies. By talking to these professionals, you should be able to obtain all the information needed to stay on the right side of the law.

Most companies use an employment manual to explain company policies to their employees. These policy manuals tell the employees what benefits they may be eligible for and when their eligibility begins. It is important that you treat all of your employees equally. You should not provide benefits for your pet employees and deny the same offering to your other employees. If you do this, you are asking for trouble. The policy manual makes it easy for you to set and maintain protocol.

MAKE PLANS FOR YOUR LATER YEARS

There is no time like the present to prepare for the future. As you plan your future, you must define the paths you want to take with your business. Will your business be handed down to your children? Will the business be run by an employee when you retire? Are you interested in selling your business at some point in the future? These are only some of the questions you should start asking yourself now.

Passing the business to your children

Passing the business along to your children is a fine way to keep your company going when you are tired of the daily grind. However, some children will have no desire to own or operate the business you spent years building. It is not that the children are ungrateful; they may just have their own dreams to fulfill.

If you have hopes of one day giving your business, or the management of it, to your children, discuss the prospect with them as soon as possible. The sooner the kids become involved in the business, the better they will be prepared to handle the responsibilities when you step down.

Don't count on your children being overly enthusiastic about taking the reins, and don't become angry with them when they want to pursue other goals. After all, you wanted to build your own business, maybe they want the same freedom. Taking over the family business can put a lot of strain on devoted children. If they do well, you might be offended that they are more capable in business than you were. If they perform poorly, they will feel they have let you down. Respect the wishes of your children, and maintain a unified family.

Allowing employees to manage your business

Allowing employees to manage your business can be a hard pill to swallow, even when it's only for a few days. If you have employees now, would you trust them to mind the store while you took a vacation? Does the thought of having someone else at the helm of your business send shivers down your back?

Getting used to putting your business into the hands of employees may take some time. If your plans call for having employees manage your business, start testing the waters now. Delegate duties to your best people and see how they handle them. When you're comfortable with their performance under your watchful eyes, take a short vacation.

You will never know how the managers will function under pressure until you let them take control. If you are standing behind them every step of the way, the managers may be nervous and not perform to their best abilities. By being too close at hand, the managers may rely on you to make the tough calls. Get away from the business and let them have a go at it. If something does go wrong, you can step back in and quickly pick up the pieces. This is the only way you are going to be able to assess fully the abilities of your chosen few.

If while you're on vacation the business runs smoothly, give the managers a little more rope. Keep testing the employees with additional responsibilities. If you have the right people, you will be able to enjoy life more and rest comfortably, knowing you have good people to back you up.

Grooming your business for sale

Grooming your business for sale is an important step towards liquidation. If you wake up one morning and decide immediately to sell your business, you are going to make mistakes. If your long-range plans call for the sale of the business, begin your preparations now. When the time comes to put the business on the auction block, you will be ready to make your best deal.

Closing the doors

Closing the doors to a business in which you have invested your life can be traumatic. You will probably feel you are throwing away a part of yourself. If shutting down is the ultimate fate for your business, prepare yourself mentally for the final days.

Reducing your workload

As an alternative to closing the doors, you might consider reducing your workload. Going into semiretirement might be the ideal answer to your problems. You can be selective in the work you do, and you can enjoy some additional income. This option is very appealing to a lot of contractors. Again, proper advance planning is the key to making your desires reality.

YOU NEED LIABILITY INSURANCE

Liability insurance is one type of insurance coverage no business can afford to be without. The extent of coverage needed will vary, but all business ventures should be protected with liability insurance. Please allow me to explain how liability insurance works.

General liability insurance protects its holder from claims arising from personal injury or property damage. When a company has a current general liability policy, all representatives of the company are typically covered under the policy when performing company business.

The cost of liability insurance will be determined by the nature of your business. Rates will be lower for someone engaged in relatively safe endeavors compared to those assessed against businesses dealing in high-risk ventures. For example, if you own a blasting company and work with explosives, your premiums will be higher than those of someone who installs interior trim molding.

Without adequate coverage against liability claims, you might lose your business and all of your other assets. Contractors are in particular need of this type of insurance. With so many possibilities for accidents on the job site, you can't afford to do business without it.

WORKER'S COMPENSATION INSURANCE IS A MUST

Worker's compensation insurance is insurance that is generally required of companies that have employees who are not close family members. The cost of this insurance, set by individual states, can be crippling, but it is a necessity for most businesses with nonfamily employees.

Worker's comp insurance benefits your employees. If employees are injured in the performance of their duties on your payroll, this insurance will help them financially. The employees may receive payment for medical expenses that are related to the injury. If employees are disabled, they may receive partial disability income from the program. Other events, such as a fatal injury, might result in similar benefits to the heirs of the employee.

The cost of worker's compensation insurance premiums is based on your company's total payroll expenses and the types of work performed by various employees. The rate for a secretary will be much lower than the rate for a roofer. Each employee is put into a job classification and rated for a degree of risk. Once the risk of injury and other factors are assessed, an estimated premium is established.

At the end of the year, the insurance carrier will conduct an audit of your company's payroll expenses to determine how much was actually paid out in payroll and to what job classifications the wages were paid. At this time the insurance company will render an accurate accounting of what is owed or due your company. Since some preliminary annual estimates are high, it is possible you will receive a refund. However, if the original estimate was low, you will have to pay the additional premiums.

Worker's comp is at best a bad experience for companies whose personnel are injured. If your company is accident prone, you will pay for it in higher premiums.

When you engage a subcontractor to work for your company, you might be held responsible for the cost of worker's compensation insurance on that sub. You can avoid this by requiring subcontractors to furnish you with a certificate of insurance before you allow them to do any work.

The certificate of insurance will come from the company that issues the insurance. Don't accept a copy of an insurance certificate that a subcontractor hands you. The policy may not be in force. When you receive the certificate of insurance, check it for coverage and expiration information. When you are satisfied that the sub has proper insurance, file the certificate for future proof of insurance.

When your insurance company audits you at the end of the year, you may need to produce certificates of insurance on all of your independent contractors. You cannot afford to let your guard down on this one. Paying premiums for insurance that subcontractors should be responsible for will cause you great grief.

Some contractors deduct money from payments due subcontractors when the subs don't carry the necessary insurance. The money is used at the end of the year when the contractors must settle up with their insurance companies. While this has been done for years, I don't recommend it. It is best to require the subcontractors to carry and provide proof of their own insurance.

RETIREMENT PLAN OPTIONS

The number of retirement plan options that exist are amazing. Whether you're looking for a plan for yourself or a plan for your employees, you will have many choices. To prove this point, let's look at some of the most common methods of building retirement capital.

Rental properties

Rental properties can be an ideal source of retirement income for yourself. Real estate is one of the best ways to keep up with the rising rates of inflation. Inflation is one of your biggest enemies when you plan for retirement. With some investments, the money earned from the investment will not amount to a hill of beans when you retire. Real estate has the edge in these circumstances because of its typical pattern of appreciation.

Rental real estate can be advantageous to you now and later. When you first buy income properties, the net rental income may not turn a profit for you, but the tax advantages can be significant. Even though the 1986 changes in tax laws dealt a deadly blow to real estate investors, there is still room to capitalize on deductions.

To make the most of your tax advantages, you must maintain an active interest in the management of your rental properties. If you are merely a passive investor, you will miss out on the bulk of the tax savings. However, being a contractor you should be well suited toward being a landlord. You have the ability and the contacts to keep maintenance costs to a minimum.

If you own rental property that breaks even, you're doing fine in your retirement plans. While you are not turning a profit, you are paying for the real estate. If you start your real estate investing early, when you retire your rental income can come to you instead of the mortgage holder.

Income-producing real estate allows you to win three ways. The first way is in the form of routine cash flow: the rents you collect will allow you to have some spending money. The second way rental properties help you is in your net worth. As you pay off your buildings, you gain equity. This equity can be used as leverage to borrow money against your properties. Since the money is borrowed, you don't have to pay taxes on it. When you have enough equity in rental real estate, you literally can live on borrowed money for the rest of your life.

The third option you have with real estate is selling it. As you have paid off the mortgages, your real estate should have increased in value. If you don't want to be an active landlord in your later years, you should be able to sell the property for a handsome profit and live off the proceeds. Since you will be selling the real estate at current prices, you will not be losing ground to inflation.

If you have the temperament and time to be a landlord, rental properties are one of the best retirement plans. You will have some problems along the way. Tenants are not always the most pleasant people and there are many laws and regulations with which you must comply. Some people simply are not cut out to be property managers. If you don't think real estate is your way to retirement riches, let's examine some other options.

Keogh plans

Keogh plans for self-employed people can get a little complicated. If you are self employed you can contribute up to 25 percent of your earnings to the fund with a maximum dollar contribution of $30,000. In reality, however, the 25 percent you are allowed to contribute is not computed on your gross earnings. Once you determine how much you are going to put in your

Keogh fund, you must subtract that amount from your gross earnings. Then, you may contribute up to 25 percent of what is left of your earnings. In effect, you can only fund 20 percent of your total earnings.

This is a little complicated, so let me give you an example. Let's say you had a great year and earned $100,000. You want to contribute $20,000 to your retirement plan. After subtracting the $20,000 from your gross earnings, you are left with $80,000. Twenty-five percent of $80,000 is $20,000, which is the maximum you can invest.

If you have employees, these plans become even more confusing. As the employer you must not only deduct the contribution to your personal plan before you arrive at the earnings figure used to factor your maximum contribution, you must also deduct the contributions you make as your part of the employees' contributions. There are other rules that apply to these plans, and as you can see, the plan can be confusing.

When you set up a Keogh plan you must name a trustee. The trustee is usually a financial institution. Before you attempt to establish and use your own Keogh plan, consult with an attorney who is familiar with the rules and regulations.

Pension plans

Pension plans for your employees must be funded in good and bad years. If you are hiring older employees, they will probably prefer a pension plan over a profit-sharing plan. Pension plans provide a consistent company contribution to the employee's retirement plan.

Pension plans are termed qualified plans. This means they meet the requirements of Section 401 of the Internal Revenue Code and qualify for favorable tax advantages. These tax advantages help you and your employees.

If you decide to use a qualified pension plan, you must cover at least 70 percent of your average employees. The features and benefits of these plans are extensive. For complete details on forming and using such a plan, consult a qualified professional.

Profit sharing

Profit-sharing plans can also be termed qualified plans. One advantage to you, as the employer, is that there is no regulation requiring you to fund the plan in bad economic years. You will need to establish a formula to identify the amount of contributions that will be made to profit-sharing plans and when contributions will be made. Many new companies prefer profit-sharing plans because there is no mandatory funding in years when a profit is not made.

Social security

Did you know that social security benefits are taxable? If an individual's adjusted gross income, tax-exempt interest, and one half of the individual's social security benefits exceed $25,000, the social security benefits can be taxed. The maximum tax is one half of the social security benefits.

Annuities

Annuities can be good retirement investments. These investments are safe, pay good interest rates, and the interest you earn is tax deferred until you cash the annuity. However, if you need access to your money early, you will have to pay a penalty for early withdrawal. If you plan to let your money work for you in the annuity for 7 to 10 years, annuities are a safe bet.

If you decide to put your money in annuities, shop around. There are a multitude of programs open to you. If you want to investigate annuity plans for your employees, talk with professionals in the field. Again, there are many options available for these programs, but there are also rules that must be followed.

Other considerations

Investing for your future can involve a variety of strategies. Bonds, art, antiques, diamonds, gold, silver, rare coins, stocks, and mutual funds are all conceivable retirement investments. Any of these forms of investments can produce a desirable rate of return. However, many of these investments require a keen knowledge of the market. For example, if you are not an experienced coin buyer, your rare coin collection may wind up being worth little more than its face value.

For most business owners, sticking to conservative investments is best for retirement. If you have some extra money you can afford to lose, you might diversify your conservative investments with some of the more exciting opportunities available. When you are betting on your golden years, play your cards carefully.

Final words

Allow me to give you some final words on retirement plans. Retirement plans for you and your employees can be quite sophisticated. With the complexity of the circumstances surrounding these plans, you should always consult experts before you make decisions.

Make yourself aware of your responsibilities to your employees. Don't assume that part-time employees are not the same as full-time employees under your benefits package. There probably are exceptions to part-time help, but don't make that assumption. Don't assume anything. Employees' rights and the law are too important to guess about. Consult professionals and maintain your integrity as an employer.

20

Plan for the future

You don't need a crystal ball to look into the future. What you need is determination, time, and skill. Time can be made and skills can be learned, but you must already possess determination. If you aren't motivated to predict the future of your business, you won't. On the other hand, if you are committed to making your business successful, you can do a fair job of projecting your business future.

Your business will face many challenges in the coming years. If you aren't prepared for these obstacles, you may not get past them. In order to be prepared, you must start making the preparations now.

Our economy runs in cycles. The businesses of most contractors are affected in some way by the real estate market. If the construction of new homes is down, most contracting fields suffer. When housing starts are up, contractors seem to thrive.

Since the economy is cyclic, you can look back into history to project the future. The clues you find may not be right on the money, but they are likely to render a clear picture of what's in store for your business. In the early 70s, the real estate market was booming. People were making and spending money. Then, in the late 70s and early 80s, the business environment took a nosedive. Interest rates soared and business production in most fields dropped. Contractors scurried to collect money due them and to find new work. Times were tough, to be sure, but many contractors made it. I was one of them.

By the mid 80s, business was good again. I was building as many as 60 homes a year, and most contractors were expanding their businesses, myself included. As time passed the economy started shifting downward. By the late 80s, the business world was reeling again. Once again, the economy was sagging and businesses were closing their doors.

Now, we are in the early 90s, and trends are pointing upward, slowly, but nevertheless upward. If you are just starting a business, what can you learn from this abbreviated lesson? You might see a trend for major slowdowns in the economy at least once every decade. You might also assume that financial failures over the last 20 years have occurred more often in the latter part of each 10-year period. Already, without much information, you can start to see that you may have to face a recession within the first 10 years of being in business.

If you look deeper into the history of the last 20 years, you will find some interesting facts. In the late 70s and early 80s, banks were quick to rise to the problems at hand. Creative financing blossomed and the business world turned itself around.

In the more recent recession of the late 80s and early 90s, banks did not rally to help. Instead, many of them closed. Business owners in this recession didn't have the high interest rates to combat, but they also didn't have willing lenders to help them with their financial battles.

With interest rates low and efforts being made to get the economy back on track for the mid and late 90s, why aren't people spending money? I believe people are afraid to spend what money they have. Many people are without jobs and the ones that have jobs don't know how long they will have them. Public confidence appears to be at an all-time low. Until confidence is restored, the rebirth of the economy will be painfully slow.

Once you are familiar with the past, you will be able to see trends as they form. You will be able to spot danger signals. These early-warning signs can be enough to save your business from financial ruin. Unless you have an astute business advisor, you will have to learn to pick up early-warning signals on your own. Read old newspapers at the library. Talk with people who have lived through tough times. Track past political performances. All these ways of looking back will help you understand the future.

LONG-RANGE PLANNING PAYS OFF

By preparing for the worst, you can handle most situations that come your way. You must focus on financial plans, management plans, and growth plans. How you plan for the future is within your grasp. To have a long-term business, you must have long-term plans.

Businesses don't change themselves. People change them, people just like you. How will you change your business? You probably don't know yet, but you had better start making plans for the changes soon.

From the field to the office

You may get tired of working in the field. What used to be fun may not be so enjoyable at an older age. The physical work that has kept you in good shape may become a bit much for you in 10 or 20 years. Some contractors plan to stay in the field until the day they retire. Many contractors anticipate hiring employees or subcontractors to do the physical work. Of the two options, I would recommend planning on hiring help. You may well get tired of working in the field before you can afford to retire.

If you know that someday you plan to bring employees or subcontractors into your business, start planning for the change now. In your spare time, if you have any, read up on human resource and management skills. The knowledge you gain now will be valuable when you enlist the help of others in your business.

Move ahead, don't stagnate

Many new business owners find something that works and stick with it. This is a good idea, so long as you don't put your business in a rut. For your business to grow and prosper, there will be times when you must step out of your comfort zone.

Since the battle for business success can get hectic, it is understandable that some people will reach a plateau and rest. I'm not suggesting that you never rest or stop to enjoy your increments of success. However, if you sit around too long, you will be left in the dust by your competitors. Being in business is not a game that allows you to win once and remain the champion. You must win regularly to stay in the game. Until you leave the business, the game is never over. There is always someone trying to cut into your market share.

Company growth

Company growth is a pattern that many business owners don't prepare for properly. These owners go about their business and add to it as volume dictates. This is a dangerous way to expand your business. Allowing your company to grow too large too fast can put you out of business. I know it may seem strange that having a bigger business might be worse than maintaining your present size, but it can.

When owners allow their companies to balloon with numerous employees, subcontractors, and jobs, management can become a serious problem. This is especially true for business owners with little management experience. The sudden wealth of quick cash flow and more jobs than you can keep up with is a company killer. The operating capital that kept your small business floating over rough waters will not be adequate to keep your new, larger business afloat. Overhead expenses will increase along with your business. These expenses may not be recovered with the pricing structure you are accustomed to using.

All in all, growing too fast can be much worse than not growing at all. If you want to expand your company, plan for the expansion. Make financial arrangements in advance, and learn the additional skills you will need to guide the business along its growth path.

Continuing education

Continuing education is a requirement for some businesses. Many licensed professionals are required by their licensing agencies to participate in continuing education. In New Hampshire, plumbers must attend an annual seminar before they can renew their licenses. In Maine, real estate brokers must complete continuing education courses to keep their licenses active. Read, attend seminars, go to classes, do whatever it takes to stay current with the changes affecting your business. If you don't keep yourself aware of the changes in your industry, you will become outdated and obsolete.

Bigger jobs

Before you venture into big jobs, make sure you can handle them. There are several factors to consider in your planning. Will you have enough money or credit to keep the big jobs and your regular work running smoothly? Do you have enough help to complete your jobs in a timely fashion? If you are required to put up a performance bond, can you? Will you be able to survive financially if the money you're anticipating from the big job is slow in coming? Do you have experience running large jobs? This line of questioning could go on for pages, but all the questions are viable ones to ask yourself. You shouldn't tackle big jobs until you are sure you can handle them.

Should you diversify?

There is no question that diversifying your company can bring you more income. But it can also cause your already successful business to get into trouble. When you split your interest into multiple fields, you are less likely to do your best at any one job. For this reason, many companies do better when they don't diversify.

In rural areas, it is sometimes necessary for a small business to fulfill many functions to survive. For example, a plumber may also do heating or electrical work. A home builder might also take on remodeling jobs. In areas with large populations, this type of diversification is not needed. Plumbers can do plumbing, builders can build, and remodelers can remodel; there is enough business to go around.

If you want to diversify, do so intelligently. There are many considerations to think about before you split your time and money into separate business interests. Most people struggle to keep one business healthy. If you get aggressive and open several business ventures, you may find you will lose them all.

Adjusting your company for change is not a task you can complete and be done with. To maintain your business, you must occasionally change your plans. Routine adjustments are normal and should be expected.

PREPARE FOR SLOW SEASONS

Many contracting businesses run on cycles. For example, a plumbing contractor in a cold climate might be flooded by calls to repair broken and frozen pipes. Lawn-care companies are not too busy cutting grass in the middle of winter. Many businesses know that they are going to have to endure ups and downs. These smart business owners make arrangements for the off season. Being prepared for slow seasons can be the difference between success or failure.

When times are tough, you may have to alter your business procedures. If you lower your prices, you will have a hard time working your prices back up to where they used to be. Lowering prices is risky business, but sometimes it is the only way to keep food on the table. If you have to lower prices to stay in business, do so with the understanding that getting prices back to normal will take time.

Depending on the nature of your work, discounts might accomplish the same goal as lowered labor rates, but with fewer long-term effects. People expect discount offers to end. Run ads offering a discount off your regular labor rates for a limited time only. This tactic will be less difficult to rebound from than lowered labor rates.

RECESSION TECHNIQUES

Hard economic times call for special techniques; I call these methods recession techniques. If you stay in business long enough, you will need to get creative to beat a bad economy. My business has been through two rough recessions. I survived both of them, and my business got stronger because of them. How could a recession make my business stronger? Hard times made me more creative, and my ideas worked, both during and after the hard times. Consequently, I had more business than ever before. Your business can weather the storms of economic slumps, but you will have to work harder and smarter.

It is well accepted that recessionary times don't stop all people from spending money. Actually, many people spend money aggressively in slow times. The people who have money know they can get their best deals when the market is down. This group of bargain hunters can provide you with plenty of business. All you have to do is find the customers and win them over. That's not too hard to do, if you put effort into it.

How do you find these spending machines during bad times? Ideally you should find and secure them before the economy sinks. You can do this by targeting your marketing and services to the right group of people. If you are already servicing the right customers when public spending drops, you will not feel as harshly the effects of the slowdown.

I think everyone would agree that word-of-mouth advertising is the best way to get good business. People talk, and if they talk favorably about your business, you will see an increase in sales. By getting in with the right customers, giving extraordinary service, and keeping your name in front of past customers, you can make yourself recession resistant.

Let me give you an example of how I did this to get past my first recession. In the late 70s and early 80s I was running a plumbing and remodeling business and had just started a home-building business. The first house I built was my own. When I started the house, home mortgage rates were around 12 percent. By the time I finished the house, just a few months later, my interest rate was 18 percent. I couldn't afford to keep the house at the higher rate, but every real estate broker I talked to told me the house couldn't be sold until rates came down.

I couldn't wait for rates to come down; I needed to sell the house fast. I ran a few ads in the local paper, but nothing happened. I decided that while the general real estate market was in terrible shape, there had to be someone out there to buy my house.

I decided to target specific groups of people. I went after investors, people who wanted country living in commuting distance to the cities, and people who wanted to have an unusual house. I ran different types of ads for each group. The ads were placed in the local paper and in more distant papers. For each group, I played up the features and benefits that would appeal to them. In less than a week, I had two people interested in the house. In less than two weeks, I had a third person and a contract for the sale of the house.

The person that bought the house was a woman who wanted to live in the country, without being too far from the city. She also fell in love with some of the special features of the home, features like a sunken bathtub, exposed beams, a front wall of glass that looked out on the forest, and the master-bedroom suite and loft.

I was prepared to sell the house for what it had cost me to build it, without charging for my labor. I thought investors would beat me down and force me to take a loss. But I didn't take a loss, I made money. I didn't make as much as I might have in a healthy market, but I made a profit and got out from under the unbearable house payments. I was able to sell the house in less than a month, even with outrageous interest rates. The point of this story is you can sell your services in any market, if you get creative and find the right people.

When I started my plumbing and remodeling business, I wanted to reach several markets. The most coveted market was upscale jobs in a specialized community. The people in this community had money and lots of it. They would literally pay $30 an hour to have their light bulbs replaced. When they remodeled a bathroom, they didn't use standard plumbing fixtures. They bought $2,500 gold faucets and expensive fixtures in high fashion colors. Most of the homes had tennis courts or swimming pools or both. The area rugs in foyers often cost thousands of dollars. It would be an understatement to say that it required a special type of contractor to work for these people.

I knew that if I could get into this market I could make strong profits and have stable work. I solicited general contractors that worked in the community. I went directly to the homeowners. It took a little time, but I got in. Once I was in, I did all the work myself. I couldn't afford to have a plumber walk on expensive rugs with muddy boots or make an off-color remark that would ruin my business reputation.

Before long I was spending three days a week, every week, working in this exclusive area. My crews took care of jobs in other areas while I continued to build up the business in this affluent neighborhood.

When the recession hit, contractors were dropping like the proverbial flies. My business suffered and struggled in most areas, but not in my special community. I am convinced that getting into that community saved my business. Even when the rest of the world around me was going downhill, my pet project was as busy as ever. The advance planning and procurement of those wealthy customers kept me going.

You may not have a rich community to tap into, but you do have special opportunities. Large corporations can keep your business afloat in difficult times. If you start doing business with the big companies before money gets tight, you will have an edge during the next economic downturn. Schools and municipal contracts are another source of constant work. If you think about the customer opportunities in your area, I'm sure you can find some that are likely to keep spending, even in bad times.

There is another problem with doing business in recessionary times. The amount of work available shrinks, and the number of people going after the work increases. People laid off from their jobs go into business for themselves. Many of these people don't pay for insurance and other business expenses that the average on-going business does. These overnight businesses operate on the principle that they are only in business until they can find another job. This type of contractor usually works cheap. For contractors carrying normal business overhead, it can be tough to compete against these pop-up contractors.

About all you can do is educate potential customers when you give estimates. Tell the customers that if they get several bids to be wary of extremely low prices. Also, advise the consumer to verify the credentials, insurance, and public standing of contractors before they do business with them. A small percentage of people will deal with people offering the lowest price, but most consumers will look further than price. If you do a good job in your sales pitch, you can win the customer's confidence and get the job, even if your price is higher.

IN CLOSING

In closing, I would like to thank you for taking this time to become a better contractor. With so much publicity about bad contractors, we all have to work to keep a shining public image. I hope your investment in this book will prove to be beneficial to you and your company. Again, thank you, and good luck in all your endeavors.

Appendix

Federal tax forms

Form **1120**
Department of the Treasury
Internal Revenue Service

U.S. Corporation Income Tax Return

For calendar year 1992 or tax year beginning ...., 1992, ending ...., 19 ...
▶ **Instructions are separate. See page 1 for Paperwork Reduction Act Notice.**

OMB No. 1545-0123

1992

A Check if a:
(1) Consolidated return (attach Form 851) ☐
(2) Personal holding co. (attach Sch. PH) ☐
(3) Personal service corp. (as defined in Temporary Regs. sec. 1.441-4T—see instructions) ☐

Use IRS label. Otherwise, please print or type.
Name
Number, street, and room or suite no. (If a P.O. box, see page 6 of instructions.)
City or town, state, and ZIP code

B Employer identification number
C Date incorporated
D Total assets (see Specific Instructions)
$

E Check applicable boxes: (1) ☐ Initial return (2) ☐ Final return (3) ☐ Change in address

Income

1a	Gross receipts or sales ___ b Less returns and allowances ___ c Bal ▶	1c	
2	Cost of goods sold (Schedule A, line 8)	2	
3	Gross profit. Subtract line 2 from line 1c	3	
4	Dividends (Schedule C, line 19)	4	
5	Interest	5	
6	Gross rents	6	
7	Gross royalties	7	
8	Capital gain net income (attach Schedule D (Form 1120))	8	
9	Net gain or (loss) from Form 4797, Part II, line 20 (attach Form 4797)	9	
10	Other income (see instructions—attach schedule)	10	
11	**Total income.** Add lines 3 through 10 ▶	11	

Deductions (See instructions for limitations on deductions.)

12	Compensation of officers (Schedule E, line 4)	12	
13a	Salaries and wages ___ b Less jobs credit ___ c Balance ▶	13c	
14	Repairs	14	
15	Bad debts	15	
16	Rents	16	
17	Taxes	17	
18	Interest	18	
19	Charitable contributions **(see instructions for 10% limitation)**	19	
20	Depreciation (attach Form 4562) ... 20		
21	Less depreciation claimed on Schedule A and elsewhere on return ... 21a	21b	
22	Depletion	22	
23	Advertising	23	
24	Pension, profit-sharing, etc., plans	24	
25	Employee benefit programs	25	
26	Other deductions (attach schedule)	26	
27	**Total deductions.** Add lines 12 through 26 ▶	27	
28	Taxable income before net operating loss deduction and special deductions. Subtract line 27 from line 11	28	
29	**Less:** a Net operating loss deduction (see instructions) ... 29a		
	b Special deductions (Schedule C, line 20) ... 29b	29c	

Tax and Payments

30	**Taxable income.** Subtract line 29c from line 28	30	
31	**Total tax** (Schedule J, line 10)	31	
32	**Payments:** a 1991 overpayment credited to 1992 ... 32a		
b	1992 estimated tax payments ... 32b		
c	Less 1992 refund applied for on Form 4466 ... 32c ( ) d Bal ▶ 32d		
e	Tax deposited with Form 7004 ... 32e		
f	Credit from regulated investment companies (attach Form 2439) ... 32f		
g	Credit for Federal tax on fuels (attach Form 4136). See instructions ... 32g	32h	
33	Estimated tax penalty (see instructions). Check if Form 2220 is attached ▶ ☐	33	
34	**Tax due.** If line 32h is smaller than the total of lines 31 and 33, enter amount owed	34	
35	**Overpayment.** If line 32h is larger than the total of lines 31 and 33, enter amount overpaid	35	
36	Enter amount of line 35 you want: **Credited to 1993 estimated tax** ▶ ___ Refunded ▶	36	

Please Sign Here

Under penalties of perjury, I declare that I have examined this return, including accompanying schedules and statements, and to the best of my knowledge and belief, it is true, correct, and complete. Declaration of preparer (other than taxpayer) is based on all information of which preparer has any knowledge.

▶ Signature of officer | Date | ▶ Title

Paid Preparer's Use Only

Preparer's signature ▶ | Date | Check if self-employed ☐ | Preparer's social security number

Firm's name (or yours if self-employed) and address ▶ | E.I. No. ▶ | ZIP code ▶

Cat. No. 11450Q

A-1a Form 1120: U.S. Corporation Income Tax Return (page 1).

Department of the Treasury—Internal Revenue Service

Schedule A — Cost of Goods Sold (See instructions.)

1	Inventory at beginning of year	1	
2	Purchases	2	
3	Cost of labor	3	
4	Additional section 263A costs (attach schedule)	4	
5	Other costs (attach schedule)	5	
6	**Total.** Add lines 1 through 5	6	
7	Inventory at end of year	7	
8	**Cost of goods sold.** Subtract line 7 from line 6. Enter here and on page 1, line 2	8	

9a Check all methods used for valuing closing inventory:

(i) ☐ Cost (ii) ☐ Lower of cost or market as described in Regulations section 1.471-4

(iii) ☐ Writedown of "subnormal" goods as described in Regulations section 1.471-2(c)

(iv) ☐ Other (Specify method used and attach explanation.) ▶ ....

b Check if the LIFO inventory method was adopted this tax year for any goods (if checked, attach Form 970) ▶ ☐

c If the LIFO inventory method was used for this tax year, enter percentage (or amounts) of closing inventory computed under LIFO ... 9c

d Do the rules of section 263A (for property produced or acquired for resale) apply to the corporation? ☐ Yes ☐ No

e Was there any change in determining quantities, cost, or valuations between opening and closing inventory? If "Yes," attach explanation ☐ Yes ☐ No

Schedule C — Dividends and Special Deductions (See instructions.)

		(a) Dividends received	(b) %	(c) Special deductions: (a) × (b)
1	Dividends from less-than-20%-owned domestic corporations that are subject to the 70% deduction (other than debt-financed stock)		70	
2	Dividends from 20%-or-more-owned domestic corporations that are subject to the 80% deduction (other than debt-financed stock)		80	
3	Dividends on debt-financed stock of domestic and foreign corporations (section 246A)		see instructions	
4	Dividends on certain preferred stock of less-than-20%-owned public utilities		41.176	
5	Dividends on certain preferred stock of 20%-or-more-owned public utilities		47.059	
6	Dividends from less-than-20%-owned foreign corporations and certain FSCs that are subject to the 70% deduction		70	
7	Dividends from 20%-or-more-owned foreign corporations and certain FSCs that are subject to the 80% deduction		80	
8	Dividends from wholly owned foreign subsidiaries subject to the 100% deduction (section 245(b))		100	
9	**Total.** Add lines 1 through 8. See instructions for limitation			
10	Dividends from domestic corporations received by a small business investment company operating under the Small Business Investment Act of 1958		100	
11	Dividends from certain FSCs that are subject to the 100% deduction (section 245(c)(1))		100	
12	Dividends from affiliated group members subject to the 100% deduction (section 243(a)(3))		100	
13	Other dividends from foreign corporations not included on lines 3, 6, 7, 8, or 11			
14	Income from controlled foreign corporations under subpart F (attach Form(s) 5471)			
15	Foreign dividend gross-up (section 78)			
16	IC-DISC and former DISC dividends not included on lines 1, 2, or 3 (section 246(d))			
17	Other dividends			
18	Deduction for dividends paid on certain preferred stock of public utilities (see instructions)			
19	**Total dividends.** Add lines 1 through 17. Enter here and on line 4, page 1 ▶			
20	**Total deductions.** Add lines 9, 10, 11, 12, and 18. Enter here and on line 29b, page 1 ▶			

Schedule E — Compensation of Officers (See instructions for line 12, page 1.)

Complete Schedule E only if total receipts (line 1a plus lines 4 through 10 on page 1, Form 1120) are $500,000 or more.

	(a) Name of officer	(b) Social security number	(c) Percent of time devoted to business	Percent of corporation stock owned: (d) Common	(e) Preferred	(f) Amount of compensation
1			%	%	%	
			%	%	%	
			%	%	%	
			%	%	%	
			%	%	%	

2	Total compensation of officers	
3	Compensation of officers claimed on Schedule A and elsewhere on return	
4	Subtract line 3 from line 2. Enter the result here and on line 12, page 1	

A-1b Form 1120: U.S. Corporation Income Tax Return (page 2).

Department of the Treasury—Internal Revenue Service

Form 1120 (1992) Page 3

Schedule J Tax Computation (See instructions.)

1 Check if the corporation is a member of a controlled group (see sections 1561 and 1563) ▶ ☐

2 If the box on line 1 is checked:

a Enter the corporation's share of the $50,000 and $25,000 taxable income bracket amounts (in that order): (i) $ (ii) $

b Enter the corporation's share of the additional 5% tax (not to exceed $11,750) ▶ $

3 Income tax. Check this box if the corporation is a qualified personal service corporation as defined in section 448(d)(2) (see instructions on page 14) ▶ ☐ — 3

4a Foreign tax credit (attach Form 1118) — 4a

b Possessions tax credit (attach Form 5735) — 4b

c Orphan drug credit (attach Form 6765) — 4c

d Credit for fuel produced from a nonconventional source — 4d

e General business credit. Enter here and check which forms are attached:
☐ Form 3800 ☐ Form 3468 ☐ Form 5884 ☐ Form 6478
☐ Form 6765 ☐ Form 8586 ☐ Form 8830 ☐ Form 8826 — 4e

f Credit for prior year minimum tax (attach Form 8827) — 4f

5 **Total credits.** Add lines 4a through 4f — 5

6 Subtract line 5 from line 3 — 6

7 Personal holding company tax (attach Schedule PH (Form 1120)) — 7

8 Recapture taxes. Check if from: ☐ Form 4255 ☐ Form 8611 — 8

9a Alternative minimum tax (attach Form 4626) — 9a

b Environmental tax (attach Form 4626) — 9b

10 **Total tax.** Add lines 6 through 9b. Enter here and on line 31, page 1 — 10

Schedule K Other Information (See instructions.) (Yes / No)

1 Check method of accounting:
a ☐ Cash b ☐ Accrual
c ☐ Other (specify) ▶

2 Refer to the list in the instructions and state the principal:
a Business activity code no. ▶
b Business activity ▶
c Product or service ▶

3 Did the corporation at the end of the tax year own, directly or indirectly, 50% or more of the voting stock of a domestic corporation? (For rules of attribution, see section 267(c).)

If "Yes," attach a schedule showing: (a) name and identifying number; (b) percentage owned; and (c) taxable income or (loss) before NOL and special deductions of such corporation for the tax year ending with or within your tax year.

4 Did any individual, partnership, corporation, estate, or trust at the end of the tax year own, directly or indirectly, 50% or more of the corporation's voting stock? (For rules of attribution, see section 267(c).) If "Yes," complete a, b, and c below

a Is the corporation a subsidiary in an affiliated group or a parent-subsidiary controlled group?

b Enter the name and identifying number of the parent corporation or other entity with 50% or more ownership ▶

c Enter percentage owned ▶

5 During this tax year, did the corporation pay dividends (other than stock dividends and distributions in exchange for stock) in excess of the corporation's current and accumulated earnings and profits? (See secs. 301 and 316.)
If "Yes," file Form 5452. If this is a consolidated return, answer here for the parent corporation and on **Form 851,** Affiliations Schedule, for each subsidiary.

6 Was the corporation a U.S. shareholder of any controlled foreign corporation? (See sections 951 and 957.)
If "Yes," attach Form 5471 for each such corporation. Enter number of Forms 5471 attached ▶

7 At any time during the 1992 calendar year, did the corporation have an interest in or a signature or other authority over a financial account in a foreign country (such as a bank account, securities account, or other financial account)?
If "Yes," the corporation may have to file Form TD F 90-22.1.
If "Yes," enter name of foreign country ▶

8 Was the corporation the grantor of, or transferor to, a foreign trust that existed during the current tax year, whether or not the corporation has any beneficial interest in it?
If "Yes," the corporation may have to file Forms 926, 3520, or 3520-A.

9 Did one foreign person at any time during the tax year own, directly or indirectly, at least 25% of: **(a)** the total voting power of all classes of stock of the corporation entitled to vote, or **(b)** the total value of all classes of stock of the corporation?
If "Yes," see page 17 of instructions and
a Enter percentage owned ▶
b Enter owner's country ▶
c The corporation may have to file Form 5472. (See page 18 for penalties that may apply.) Enter number of Forms 5472 attached ▶

10 Check this box if the corporation issued publicly offered debt instruments with original issue discount ▶ ☐
If so, the corporation may have to file Form 8281.

11 Enter the amount of tax-exempt interest received or accrued during the tax year ▶ $

12 If there were 35 or fewer shareholders at the end of the tax year, enter the number ▶

13 If the corporation has an NOL for the tax year and is electing under sec. 172(b)(3) to forego the carryback period, check here ▶ ☐

A-1c Form 1120: U.S. Corporation Income Tax Return (page 3).

Department of the Treasury—Internal Revenue Service

Form 1120 (1992) Page 4

Schedule L Balance Sheets

	Beginning of tax year		End of tax year	
Assets	(a)	(b)	(c)	(d)
1 Cash				
2a Trade notes and accounts receivable				
b Less allowance for bad debts	( )		( )	
3 Inventories				
4 U.S. government obligations				
5 Tax-exempt securities (see instructions)				
6 Other current assets (attach schedule)				
7 Loans to stockholders				
8 Mortgage and real estate loans				
9 Other investments (attach schedule)				
10a Buildings and other depreciable assets				
b Less accumulated depreciation	( )		( )	
11a Depletable assets				
b Less accumulated depletion	( )		( )	
12 Land (net of any amortization)				
13a Intangible assets (amortizable only)				
b Less accumulated amortization	( )		( )	
14 Other assets (attach schedule)				
15 Total assets				
Liabilities and Stockholders' Equity				
16 Accounts payable				
17 Mortgages, notes, bonds payable in less than 1 year				
18 Other current liabilities (attach schedule)				
19 Loans from stockholders				
20 Mortgages, notes, bonds payable in 1 year or more				
21 Other liabilities (attach schedule)				
22 Capital stock: a Preferred stock				
b Common stock				
23 Paid-in or capital surplus				
24 Retained earnings—Appropriated (attach schedule)				
25 Retained earnings—Unappropriated				
26 Less cost of treasury stock		( )		( )
27 Total liabilities and stockholders' equity				

Note: *You are not required to complete Schedules M-1 and M-2 below if the total assets on line 15, column (d) of Schedule L are less than $25,000.*

Schedule M-1 Reconciliation of Income (Loss) per Books With Income per Return (See instructions.)

1 Net income (loss) per books

2 Federal income tax

3 Excess of capital losses over capital gains

4 Income subject to tax not recorded on books this year (itemize):

5 Expenses recorded on books this year not deducted on this return (itemize):

a Depreciation . . . . $

b Contributions carryover $

c Travel and entertainment $

6 Add lines 1 through 5

7 Income recorded on books this year not included on this return (itemize):

Tax-exempt interest $

8 Deductions on this return not charged against book income this year (itemize):

a Depreciation . . . . $

b Contributions carryover $

9 Add lines 7 and 8

10 Income (line 28, page 1)—line 6 less line 9

Schedule M-2 Analysis of Unappropriated Retained Earnings per Books (Line 25, Schedule L)

1 Balance at beginning of year

2 Net income (loss) per books

3 Other increases (itemize):

4 Add lines 1, 2, and 3

5 Distributions: a Cash

b Stock

c Property

6 Other decreases (itemize):

7 Add lines 5 and 6

8 Balance at end of year (line 4 less line 7)

A-1d Form 1120: U.S. Corporation Income Tax Return (page 4).

Department of the Treasury—Internal Revenue Service

Form **1120S**
Department of the Treasury
Internal Revenue Service

U.S. Income Tax Return for an S Corporation

For calendar year 1992, or tax year beginning ...., 1992, and ending ...., 19 .....
▶ See separate instructions.

OMB No. 1545-0130

1992

	Use IRS label. Otherwise, please print or type.	
A Date of election as an S corporation	Name	C Employer identification number
	Number, street, and room or suite no. (If a P.O. box, see page 8 of the instructions.)	D Date incorporated
B Business code no. (see Specific Instructions)	City or town, state, and ZIP code	E Total assets (see Specific Instructions) $

F Check applicable boxes: (1) ☐ Initial return (2) ☐ Final return (3) ☐ Change in address (4) ☐ Amended return

G Check this box if this S corporation is subject to the consolidated audit procedures of sections 6241 through 6245 (see instructions before checking this box) . ▶ ☐

H Enter number of shareholders in the corporation at end of the tax year . . . ▶

Caution: *Include **only** trade or business income and expenses on lines 1a through 21. See the instructions for more information.*

Income	1a	Gross receipts or sales	b Less returns and allowances	c Bal ▶	1c
	2	Cost of goods sold (Schedule A, line 8)			2
	3	Gross profit. Subtract line 2 from line 1c			3
	4	Net gain (loss) from Form 4797, Part II, line 20 *(attach Form 4797)*			4
	5	Other income (loss) (see instructions) *(attach schedule)*			5
	6	**Total income (loss).** Combine lines 3 through 5 ▶			6
Deductions (See instructions for limitations.)	7	Compensation of officers			7
	8a	Salaries and wages	b Less jobs credit	c Bal ▶	8c
	9	Repairs			9
	10	Bad debts			10
	11	Rents			11
	12	Taxes			12
	13	Interest			13
	14a	Depreciation (see instructions)	14a		
	b	Depreciation claimed on Schedule A and elsewhere on return	14b		
	c	Subtract line 14b from line 14a			14c
	15	Depletion **(Do not deduct oil and gas depletion.)**			15
	16	Advertising			16
	17	Pension, profit-sharing, etc., plans			17
	18	Employee benefit programs			18
	19	Other deductions (see instructions) *(attach schedule)*			19
	20	**Total deductions.** Add lines 7 through 19 ▶			20
	21	Ordinary income (loss) from trade or business activities. Subtract line 20 from line 6			21
Tax and Payments	22	**Tax:**			
	a	Excess net passive income tax *(attach schedule)*	22a		
	b	Tax from Schedule D (Form 1120S)	22b		
	c	Add lines 22a and 22b (see instructions for additional taxes)			22c
	23	**Payments:**			
	a	1992 estimated tax payments	23a		
	b	Tax deposited with Form 7004	23b		
	c	Credit for Federal tax paid on fuels *(attach Form 4136)*	23c		
	d	Add lines 23a through 23c			23d
	24	Estimated tax penalty (see instructions). Check if Form 2220 is attached ▶ ☐			24
	25	**Tax due.** If the total of lines 22c and 24 is larger than line 23d, enter amount owed. See instructions for depositary method of payment ▶			25
	26	**Overpayment.** If line 23d is larger than the total of lines 22c and 24, enter amount overpaid ▶			26
	27	Enter amount of line 26 you want: **Credited to 1993 estimated tax** ▶	**Refunded** ▶		27

Please Sign Here

Under penalties of perjury, I declare that I have examined this return, including accompanying schedules and statements, and to the best of my knowledge and belief, it is true, correct, and complete. Declaration of preparer (other than taxpayer) is based on all information of which preparer has any knowledge.

▶ Signature of officer | Date | ▶ Title

Paid Preparer's Use Only

Preparer's signature ▶	Date	Check if self-employed ▶ ☐	Preparer's social security number
Firm's name (or yours if self-employed) and address ▶		E.I. No. ▶	
		ZIP code ▶	

For Paperwork Reduction Act Notice, see page 1 of separate instructions. Cat. No. 11510H Form **1120S** (1992)

A-2a Form 1120S: U.S. Income Tax Return for an S Corporation (page 1).

Department of the Treasury—Internal Revenue Service

Form 1120S (1992) Page **2**

Schedule A Cost of Goods Sold (See instructions.)

1	Inventory at beginning of year	1	
2	Purchases	2	
3	Cost of labor	3	
4	Additional section 263A costs (see instructions) *(attach schedule)*	4	
5	Other costs *(attach schedule)*.	5	
6	**Total.** Add lines 1 through 5	6	
7	Inventory at end of year	7	
8	**Cost of goods sold.** Subtract line 7 from line 6. Enter here and on page 1, line 2	8	

9a Check all methods used for valuing closing inventory:

(i) ☐ Cost

(ii) ☐ Lower of cost or market as described in Regulations section 1.471-4

(iii) ☐ Writedown of "subnormal" goods as described in Regulations section 1.471-2(c)

(iv) ☐ Other (specify method used and attach explanation) ▶ ....

b Check if the LIFO inventory method was adopted this tax year for any goods *(if checked, attach Form 970)*. ▶ ☐

c If the LIFO inventory method was used for this tax year, enter percentage (or amounts) of closing inventory computed under LIFO . . . **9c**

d Do the rules of section 263A (for property produced or acquired for resale) apply to the corporation? ☐ Yes ☐ No

e Was there any change in determining quantities, cost, or valuations between opening and closing inventory? ☐ Yes ☐ No
If "Yes," attach explanation.

Schedule B Other Information

		Yes	No
1	Check method of accounting: **(a)** ☐ Cash **(b)** ☐ Accrual **(c)** ☐ Other (specify) ▶		
2	Refer to the list in the instructions and state the corporation's principal: **(a)** Business activity ▶ .... **(b)** Product or service ▶		
3	Did the corporation at the end of the tax year own, directly or indirectly, 50% or more of the voting stock of a domestic corporation? (For rules of attribution, see section 267(c).) If "Yes," attach a schedule showing: **(a)** name, address, and employer identification number and **(b)** percentage owned.		
4	Was the corporation a member of a controlled group subject to the provisions of section 1561?		
5	At any time during calendar year 1992, did the corporation have an interest in or a signature or other authority over a financial account in a foreign country (such as a bank account, securities account, or other financial account)? (See instructions for exceptions and filing requirements for form TD F 90-22.1.)		
	If "Yes," enter the name of the foreign country ▶		
6	Was the corporation the grantor of, or transferor to, a foreign trust that existed during the current tax year, whether or not the corporation has any beneficial interest in it? If "Yes," the corporation may have to file Forms 3520, 3520-A, or 926		
7	Check this box if the corporation has filed or is required to file **Form 8264,** Application for Registration of a Tax Shelter ▶ ☐		
8	Check this box if the corporation issued publicly offered debt instruments with original issue discount ▶ ☐ If so, the corporation may have to file **Form 8281,** Information Return for Publicly Offered Original Issue Discount Instruments.		
9	If the corporation: **(a)** filed its election to be an S corporation after 1986, **(b)** was a C corporation before it elected to be an S corporation **or** the corporation acquired an asset with a basis determined by reference to its basis (or the basis of any other property) in the hands of a C corporation, and **(c)** has net unrealized built-in gain (defined in section 1374(d)(1)) in excess of the net recognized built-in gain from prior years, enter the net unrealized built-in gain reduced by net recognized built-in gain from prior years (see instructions) ▶ $		
10	Check this box if the corporation had subchapter C earnings and profits at the close of the tax year (see instructions) ▶ ☐		
11	Was this corporation in operation at the end of 1992?		
12	How many months in 1992 was this corporation in operation?		

Designation of Tax Matters Person (See instructions.)

Enter below the shareholder designated as the tax matters person (TMP) for the tax year of this return:

Name of designated TMP ▶ Identifying number of TMP ▶

Address of designated TMP ▶

A-2b Form 1120S: U.S. Income Tax Return for an S Corporation (page 2).

Department of the Treasury—Internal Revenue Service

Schedule K Shareholders' Shares of Income, Credits, Deductions, etc.

	(a) Pro rata share items		(b) Total amount
Income (Loss)	1 Ordinary income (loss) from trade or business activities (page 1, line 21)	1	
	2 Net income (loss) from rental real estate activities *(attach Form 8825)*	2	
	3a Gross income from other rental activities — 3a		
	b Expenses from other rental activities *(attach schedule)* — 3b		
	c Net income (loss) from other rental activities. Subtract line 3b from line 3a	3c	
	4 Portfolio income (loss):		
	a Interest income	4a	
	b Dividend income	4b	
	c Royalty income	4c	
	d Net short-term capital gain (loss) *(attach Schedule D (Form 1120S))*	4d	
	e Net long-term capital gain (loss) *(attach Schedule D (Form 1120S))*	4e	
	f Other portfolio income (loss) *(attach schedule)*	4f	
	5 Net gain (loss) under section 1231 (other than due to casualty or theft) *(attach Form 4797)*	5	
	6 Other income (loss) *(attach schedule)*	6	
Deductions	7 Charitable contributions (see instructions) *(attach schedule)*	7	
	8 Section 179 expense deduction *(attach Form 4562)*	8	
	9 Deductions related to portfolio income (loss) (see instructions) (itemize)	9	
	10 Other deductions *(attach schedule)*	10	
Investment Interest	11a Interest expense on investment debts	11a	
	b (1) Investment income included on lines 4a through 4f above	11b(1)	
	(2) Investment expenses included on line 9 above	11b(2)	
Credits	12a Credit for alcohol used as a fuel *(attach Form 6478)*	12a	
	b Low-income housing credit (see instructions):		
	(1) From partnerships to which section 42(j)(5) applies for property placed in service before 1990	12b(1)	
	(2) Other than on line 12b(1) for property placed in service before 1990	12b(2)	
	(3) From partnerships to which section 42(j)(5) applies for property placed in service after 1989	12b(3)	
	(4) Other than on line 12b(3) for property placed in service after 1989	12b(4)	
	c Qualified rehabilitation expenditures related to rental real estate activities *(attach Form 3468)*	12c	
	d Credits (other than credits shown on lines 12b and 12c) related to rental real estate activities (see instructions)	12d	
	e Credits related to other rental activities (see instructions)	12e	
	13 Other credits (see instructions)	13	
Adjustments and Tax Preference Items	14a Depreciation adjustment on property placed in service after 1986	14a	
	b Adjusted gain or loss	14b	
	c Depletion (other than oil and gas)	14c	
	d (1) Gross income from oil, gas, or geothermal properties	14d(1)	
	(2) Deductions allocable to oil, gas, or geothermal properties	14d(2)	
	e Other adjustments and tax preference items *(attach schedule)*	14e	
Foreign Taxes	15a Type of income ▶		
	b Name of foreign country or U.S. possession ▶		
	c Total gross income from sources outside the United States *(attach schedule)*	15c	
	d Total applicable deductions and losses *(attach schedule)*	15d	
	e Total foreign taxes (check one): ▶ ☐ Paid ☐ Accrued	15e	
	f Reduction in taxes available for credit *(attach schedule)*	15f	
	g Other foreign tax information *(attach schedule)*	15g	
Other	16a Total expenditures to which a section 59(e) election may apply	16a	
	b Type of expenditures ▶		
	17 Tax-exempt interest income	17	
	18 Other tax-exempt income	18	
	19 Nondeductible expenses	19	
	20 Total property distributions (including cash) other than dividends reported on line 22 below	20	
	21 Other items and amounts required to be reported separately to shareholders (see instructions) *(attach schedule)*		
	22 Total dividend distributions paid from accumulated earnings and profits	22	
	23 **Income (loss).** (Required only if Schedule M-1 must be completed.) Combine lines 1 through 6 in column (b). From the result, subtract the sum of lines 7 through 11a, 15e, and 16a	23	

A-2c Form 1120S: U.S. Income Tax Return for an S Corporation (page 3).

Department of the Treasury—Internal Revenue Service

Form 1120S (1992) Page 4

Schedule L — Balance Sheets

Assets	Beginning of tax year (a)	(b)	End of tax year (c)	(d)
1 Cash				
2a Trade notes and accounts receivable				
b Less allowance for bad debts				
3 Inventories				
4 U.S. Government obligations				
5 Tax-exempt securities				
6 Other current assets *(attach schedule)*				
7 Loans to shareholders				
8 Mortgage and real estate loans				
9 Other investments *(attach schedule)*				
10a Buildings and other depreciable assets				
b Less accumulated depreciation				
11a Depletable assets				
b Less accumulated depletion				
12 Land (net of any amortization)				
13a Intangible assets (amortizable only)				
b Less accumulated amortization				
14 Other assets *(attach schedule)*				
15 Total assets				
Liabilities and Shareholders' Equity				
16 Accounts payable				
17 Mortgages, notes, bonds payable in less than 1 year				
18 Other current liabilities *(attach schedule)*				
19 Loans from shareholders				
20 Mortgages, notes, bonds payable in 1 year or more				
21 Other liabilities *(attach schedule)*				
22 Capital stock				
23 Paid-in or capital surplus				
24 Retained earnings				
25 Less cost of treasury stock		( )		( )
26 Total liabilities and shareholders' equity				

Schedule M-1 — Reconciliation of Income (Loss) per Books With Income (Loss) per Return (You are not required to complete this schedule if the total assets on line 15, column (d), of Schedule L are less than $25,000.)

1 Net income (loss) per books		5 Income recorded on books this year not included on Schedule K, lines 1 through 6 (itemize):	
2 Income included on Schedule K, lines 1 through 6, not recorded on books this year (itemize):		a Tax-exempt interest $	
3 Expenses recorded on books this year not included on Schedule K, lines 1 through 11a, 15e, and 16a (itemize):		6 Deductions included on Schedule K, lines 1 through 11a, 15e, and 16a, not charged against book income this year (itemize):	
a Depreciation $		a Depreciation $	
b Travel and entertainment $			
		7 Add lines 5 and 6	
4 Add lines 1 through 3		8 Income (loss) (Schedule K, line 23). Line 4 less line 7	

Schedule M-2 — Analysis of Accumulated Adjustments Account, Other Adjustments Account, and Shareholders' Undistributed Taxable Income Previously Taxed (See instructions.)

	(a) Accumulated adjustments account	(b) Other adjustments account	(c) Shareholders' undistributed taxable income previously taxed
1 Balance at beginning of tax year			
2 Ordinary income from page 1, line 21			
3 Other additions			
4 Loss from page 1, line 21	( )		
5 Other reductions	( )	( )	
6 Combine lines 1 through 5			
7 Distributions other than dividend distributions			
8 Balance at end of tax year. Subtract line 7 from line 6			

*U.S. Government Printing Office: 1992 — 315-273

A-2d Form 1120S: U.S. Income Tax Return for an S Corporation (page 4).

Department of the Treasury—Internal Revenue Service

Form **1065** | **U.S. Partnership Return of Income** | OMB No. 1545-0099

Department of the Treasury
Internal Revenue Service

For calendar year 1992, or tax year beginning .... , 1992, and ending .... , 19 .... .
▶ **See separate instructions.**

1992

	Use the IRS label. Otherwise, please print or type.		
A Principal business activity		Name of partnership	D Employer identification number
B Principal product or service		Number, street, and room or suite no. (If a P.O. box, see page 9 of the instructions.)	E Date business started
C Business code number		City or town, state, and ZIP code	F Total assets (see Specific Instructions) $

G Check applicable boxes: (1) ☐ Initial return (2) ☐ Final return (3) ☐ Change in address (4) ☐ Amended return

H Check accounting method: (1) ☐ Cash (2) ☐ Accrual (3) ☐ Other (specify) ▶ ....

I Number of partners in this partnership . . . ▶ ....

Caution: *Include **only** trade or business income and expenses on lines 1a through 22 below. See the instructions for more information.*

Income

1a Gross receipts or sales	1a		
b Less returns and allowances	1b	1c	
2 Cost of goods sold (Schedule A, line 8)		2	
3 Gross profit. Subtract line 2 from line 1c		3	
4 Ordinary income (loss) from other partnerships and fiduciaries *(attach schedule)*		4	
5 Net farm profit (loss) *(attach Schedule F (Form 1040))*		5	
6 Net gain (loss) from Form 4797, Part II, line 20		6	
7 Other income (loss) (see instructions) *(attach schedule)*		7	
8 **Total income (loss).** Combine lines 3 through 7		8	

Deductions (see instructions for limitations)

9a Salaries and wages (other than to partners)	9a		
b Less jobs credit	9b	9c	
10 Guaranteed payments to partners		10	
11 Repairs		11	
12 Bad debts		12	
13 Rent		13	
14 Taxes		14	
15 Interest		15	
16a Depreciation (see instructions)	16a		
b Less depreciation reported on Schedule A and elsewhere on return	16b	16c	
17 Depletion **(Do not deduct oil and gas depletion.)**		17	
18 Retirement plans, etc.		18	
19 Employee benefit programs		19	
20 Other deductions *(attach schedule)*		20	
21 **Total deductions.** Add the amounts shown in the far right column for lines 9c through 20		21	
22 **Ordinary income (loss)** from trade or business activities. Subtract line 21 from line 8		22	

Please Sign Here

Under penalties of perjury, I declare that I have examined this return, including accompanying schedules and statements, and to the best of my knowledge and belief, it is true, correct, and complete. Declaration of preparer (other than general partner) is based on all information of which preparer has any knowledge.

▶ Signature of general partner ▶ Date

Paid Preparer's Use Only

Preparer's signature ▶	Date	Check if self-employed ▶ ☐	Preparer's social security no.
Firm's name (or yours if self-employed) and address ▶		E.I. No. ▶	
		ZIP code ▶	

For Paperwork Reduction Act Notice, see page 1 of separate instructions. Cat. No. 11390Z Form **1065** (1992)

A-3a Form 1065: U.S. Partnership Return of Income (page 1).

Department of the Treasury—Internal Revenue Service

Form 1065 (1992) Page **2**

Schedule A Cost of Goods Sold

1	Inventory at beginning of year	1	
2	Purchases less cost of items withdrawn for personal use	2	
3	Cost of labor	3	
4	Additional section 263A costs (see instructions) *(attach schedule)*	4	
5	Other costs *(attach schedule)*	5	
6	**Total.** Add lines 1 through 5	6	
7	Inventory at end of year	7	
8	**Cost of goods sold.** Subtract line 7 from line 6. Enter here and on page 1, line 2	8	

9a Check all methods used for valuing closing inventory:

- **(i)** ☐ Cost
- **(ii)** ☐ Lower of cost or market as described in Regulations section 1.471-4
- **(iii)** ☐ Writedown of "subnormal" goods as described in Regulations section 1.471-2(c)
- **(iv)** ☐ Other (specify method used and attach explanation) ▶

b Check this box if the LIFO inventory method was adopted this tax year for any goods *(if checked, attach Form 970)* ▶ ☐

c Do the rules of section 263A (for property produced or acquired for resale) apply to the partnership? ☐ **Yes** ☐ **No**

d Was there any change in determining quantities, cost, or valuations between opening and closing inventory? ☐ **Yes** ☐ **No**
If "Yes," attach explanation.

Schedule B Other Information

		Yes	No
1	Is this partnership a limited partnership?		
2	Are any partners in this partnership also partnerships?		
3	Is this partnership a partner in another partnership?		
4	Is this partnership subject to the consolidated audit procedures of sections 6221 through 6233? If "Yes," see **Designation of Tax Matters Partner** below		
5	Does this partnership meet **ALL THREE** of the following requirements?		
a	The partnership's total receipts for the tax year were less than $250,000;		
b	The partnership's total assets at the end of the tax year were less than $250,000; **AND**		
c	Schedules K-1 are filed with the return and furnished to the partners on or before the due date (including extensions) for the partnership return. If "Yes," the partnership is not required to complete Schedules L, M-1, and M-2; Item F on page 1 of Form 1065; or Item J on Schedule K-1		
6	Does this partnership have any foreign partners?		
7	Is this partnership a publicly traded partnership as defined in section 469(k)(2)?		
8	Has this partnership filed, or is it required to file, **Form 8264,** Application for Registration of a Tax Shelter?		
9	At any time during calendar year 1992, did the partnership have an interest in or a signature or other authority over a financial account in a foreign country (such as a bank account, securities account, or other financial account)? (See the instructions for exceptions and filing requirements for form TD F 90-22.1.) If "Yes," enter the name of the foreign country. ▶		
10	Was the partnership the grantor of, or transferor to, a foreign trust that existed during the current tax year, whether or not the partnership or any partner has any beneficial interest in it? If "Yes," you may have to file Forms 3520, 3520-A, or 926		
11	Was there a distribution of property or a transfer (e.g., by sale or death) of a partnership interest during the tax year? If "Yes," you may elect to adjust the basis of the partnership's assets under section 754 by attaching the statement described under **Elections** on page 5 of the instructions		
12	Was this partnership in operation at the end of 1992?		
13	How many months in 1992 was this partnership actively operated? ▶		

Designation of Tax Matters Partner (See instructions.)
Enter below the general partner designated as the tax matters partner (TMP) for the tax year of this return:

Name of designated TMP ▶ Identifying number of TMP ▶

Address of designated TMP ▶

A-3b Form 1065: U.S. Partnership Return of Income (page 2).

Department of the Treasury—Internal Revenue Service

Form 1065 (1992) Page 3

Schedule K **Partners' Shares of Income, Credits, Deductions, Etc.**

	(a) Distributive share items		(b) Total amount
Income (Loss)	1 Ordinary income (loss) from trade or business activities (page 1, line 22)	1	
	2 Net income (loss) from rental real estate activities *(attach Form 8825)*	2	
	3a Gross income from other rental activities — 3a		
	b Expenses from other rental activities*(attach schedule)* — 3b		
	c Net income (loss) from other rental activities. Subtract line 3b from line 3a	3c	
	4 Portfolio income (loss) (see instructions): **a** Interest income	4a	
	b Dividend income	4b	
	c Royalty income	4c	
	d Net short-term capital gain (loss) *(attach Schedule D (Form 1065))*	4d	
	e Net long-term capital gain (loss) *(attach Schedule D (Form 1065))*	4e	
	f Other portfolio income (loss) *(attach schedule)*	4f	
	5 Guaranteed payments to partners	5	
	6 Net gain (loss) under section 1231 (other than due to casualty or theft) *(attach Form 4797)*	6	
	7 Other income (loss) *(attach schedule)*	7	
Deductions	8 Charitable contributions (see instructions) *(attach schedule)*	8	
	9 Section 179 expense deduction *(attach Form 4562)*	9	
	10 Deductions related to portfolio income (see instructions) (itemize)	10	
	11 Other deductions *(attach schedule)*	11	
Investment Interest	12a Interest expense on investment debts	12a	
	b (1) Investment income included on lines 4a through 4f above	12b(1)	
	(2) Investment expenses included on line 10 above	12b(2)	
Credits	13a Credit for income tax withheld	13a	
	b Low-income housing credit (see instructions):		
	(1) From partnerships to which section 42(j)(5) applies for property placed in service before 1990	13b(1)	
	(2) Other than on line 13b(1) for property placed in service before 1990	13b(2)	
	(3) From partnerships to which section 42(j)(5) applies for property placed in service after 1989	13b(3)	
	(4) Other than on line 13b(3) for property placed in service after 1989	13b(4)	
	c Qualified rehabilitation expenditures related to rental real estate activities *(attach Form 3468)*	13c	
	d Credits (other than credits shown on lines 13b and 13c) related to rental real estate activities (see instructions)	13d	
	e Credits related to other rental activities (see instructions)	13e	
	14 Other credits (see instructions)	14	
Self-Employment	15a Net earnings (loss) from self-employment	15a	
	b Gross farming or fishing income	15b	
	c Gross nonfarm income	15c	
Adjustments and Tax Preference Items	16a Depreciation adjustment on property placed in service after 1986	16a	
	b Adjusted gain or loss	16b	
	c Depletion (other than oil and gas)	16c	
	d (1) Gross income from oil, gas, and geothermal properties	16d(1)	
	(2) Deductions allocable to oil, gas, and geothermal properties	16d(2)	
	e Other adjustments and tax preference items *(attach schedule)*	16e	
Foreign Taxes	17a Type of income ▶ .... **b** Foreign country or U.S. possession ▶		
	c Total gross income from sources outside the United States *(attach schedule)*.	17c	
	d Total applicable deductions and losses *(attach schedule)*	17d	
	e Total foreign taxes (check one): ▶ ☐ Paid ☐ Accrued	17e	
	f Reduction in taxes available for credit *(attach schedule)*	17f	
	g Other foreign tax information *(attach schedule)*	17g	
Other	18a Total expenditures to which a section 59(e) election may apply	18a	
	b Type of expenditures ▶		
	19 Tax-exempt interest income	19	
	20 Other tax-exempt income	20	
	21 Nondeductible expenses	21	
	22 Other items and amounts required to be reported separately to partners (see instructions) *(attach schedule)*		
Analysis	23a Income (loss). Combine lines 1 through 7 in column (b). From the result, subtract the sum of lines 8 through 12a, 17e, and 18a	23a	

b Analysis by type of partner:	(a) Corporate	(b) Individual i. Active	(b) Individual ii. Passive	(c) Partnership	(d) Exempt organization	(e) Nominee/Other
(1) General partners						
(2) Limited partners						

A-3c Form 1065: U.S. Partnership Return of Income (page 3).

Department of the Treasury—Internal Revenue Service

Form 1065 (1992) Page 4

Caution: *If Question 5 of Schedule B is answered "Yes," the partnership is not required to complete Schedules L, M-1, and M-2.*

Schedule L — Balance Sheets

Assets	Beginning of tax year (a)	(b)	End of tax year (c)	(d)
1 Cash				
2a Trade notes and accounts receivable				
b Less allowance for bad debts				
3 Inventories				
4 U.S. government obligations				
5 Tax-exempt securities				
6 Other current assets *(attach schedule)*				
7 Mortgage and real estate loans				
8 Other investments *(attach schedule)*				
9a Buildings and other depreciable assets				
b Less accumulated depreciation				
10a Depletable assets				
b Less accumulated depletion				
11 Land (net of any amortization)				
12a Intangible assets (amortizable only)				
b Less accumulated amortization				
13 Other assets *(attach schedule)*				
14 **Total** assets				
Liabilities and Capital				
15 Accounts payable				
16 Mortgages, notes, bonds payable in less than 1 year				
17 Other current liabilities *(attach schedule)*				
18 All nonrecourse loans				
19 Mortgages, notes, bonds payable in 1 year or more				
20 Other liabilities *(attach schedule)*				
21 Partners' capital accounts				
22 **Total** liabilities and capital				

Schedule M-1 — Reconciliation of Income (Loss) per Books With Income (Loss) per Return (see instructions)

1 Net income (loss) per books

2 Income included on Schedule K, lines 1 through 4, 6, and 7, not recorded on books this year (itemize):

3 Guaranteed payments (other than health insurance)

4 Expenses recorded on books this year not included on Schedule K, lines 1 through 12a, 17e, and 18a (itemize):

a Depreciation $

b Travel and entertainment $

5 Total of lines 1 through 4

6 Income recorded on books this year not included on Schedule K, lines 1 through 7 (itemize):

a Tax-exempt interest $

7 Deductions included on Schedule K, lines 1 through 12a, 17e, and 18a, not charged against book income this year (itemize):

a Depreciation $

8 Total of lines 6 and 7

9 Income (loss) (Schedule K, line 23a). Subtract line 8 from line 5

Schedule M-2 — Analysis of Partners' Capital Accounts

1 Balance at beginning of year

2 Capital contributed during year

3 Net income (loss) per books

4 Other increases (itemize):

5 Total of lines 1 through 4

6 Distributions: a Cash

b Property

7 Other decreases (itemize):

8 Total of lines 6 and 7

9 Balance at end of year. Subtract line 8 from line 5

*U.S. Government Printing Office: 1992 — 315-219

A-3d Form 1065: U.S. Partnership Return of Income (page 4).

Department of the Treasury—Internal Revenue Service

SCHEDULE C (Form 1040)

Department of the Treasury
Internal Revenue Service (0)

Profit or Loss From Business

(Sole Proprietorship)

▶ Partnerships, joint ventures, etc., must file Form 1065.

▶ Attach to Form 1040 or Form 1041. ▶ See Instructions for Schedule C (Form 1040).

OMB No. 1545-0074

1992

Attachment Sequence No. **09**

Name of proprietor	**Social security number (SSN)**
A Principal business or profession, including product or service (see page C-1)	**B** Enter principal business code (from page 2) ▶
C Business name	**D Employer ID number (Not SSN)**

E Business address (including suite or room no.) ▶ ....
City, town or post office, state, and ZIP code

F Accounting method: **(1)** ☐ Cash **(2)** ☐ Accrual **(3)** ☐ Other (specify) ▶ ....

G Method(s) used to value closing inventory: **(1)** ☐ Cost **(2)** ☐ Lower of cost or market **(3)** ☐ Other (attach explanation) **(4)** ☐ Does not apply (if checked, skip line H)

		Yes	No
H	Was there any change in determining quantities, costs, or valuations between opening and closing inventory? If "Yes," attach explanation		
I	Did you "materially participate" in the operation of this business during 1992? If "No," see page C-2 for limitations on losses		
J	Was this business in operation at the end of 1992?		
K	How many months was this business in operation during 1992? ▶		
L	If this is the first Schedule C filed for this business, check here ▶ ☐		

Part I Income

1	Gross receipts or sales. **Caution:** *If this income was reported to you on Form W-2 and the "Statutory employee" box on that form was checked, see page C-2 and check here* ▶ ☐	1	
2	Returns and allowances	2	
3	Subtract line 2 from line 1	3	
4	Cost of goods sold (from line 40 on page 2)	4	
5	**Gross profit.** Subtract line 4 from line 3	5	
6	Other income, including Federal and state gasoline or fuel tax credit or refund (see page C-2)	6	
7	**Gross income.** Add lines 5 and 6 ▶	7	

Part II Expenses (Caution: *Do not* enter expenses for business use of your home on lines 8–27. Instead, see line 30.)

8	Advertising	8		21	Repairs and maintenance	21	
9	Bad debts from sales or services (see page C-3)	9		22	Supplies (not included in Part III)	22	
10	Car and truck expenses (see page C-3—also attach **Form 4562**)	10		23	Taxes and licenses	23	
11	Commissions and fees	11		24	Travel, meals, and entertainment:		
12	Depletion	12		a	Travel	24a	
13	Depreciation and section 179 expense deduction (not included in Part III) (see page C-3)	13		b	Meals and entertainment		
14	Employee benefit programs (other than on line 19)	14		c	Enter 20% of line 24b subject to limitations (see page C-4)		
15	Insurance (other than health)	15		d	Subtract line 24c from line 24b	24d	
16	Interest:			25	Utilities	25	
a	Mortgage (paid to banks, etc.)	16a		26	Wages (less jobs credit)	26	
b	Other	16b		27a	Other expenses (**list type and amount**):		
17	Legal and professional services	17					
18	Office expense	18					
19	Pension and profit-sharing plans	19					
20	Rent or lease (see page C-4):						
a	Vehicles, machinery, and equipment	20a					
b	Other business property	20b		27b	Total other expenses	27b	

28	**Total expenses** before expenses for business use of home. Add lines 8 through 27b in columns ▶	28	
29	Tentative profit (loss). Subtract line 28 from line 7	29	
30	Expenses for business use of your home. Attach **Form 8829**	30	
31	**Net profit or (loss).** Subtract line 30 from line 29. If a profit, enter here and on Form 1040, line 12. Also, enter the net profit on Schedule SE, line 2 (statutory employees, see page C-5). If a loss, you MUST go on to line 32 (fiduciaries, see page C-5)	31	
32	If you have a loss, you MUST check the box that describes your investment in this activity (see page C-5). If you checked 32a, enter the loss on Form 1040, line 12, and Schedule SE, line 2 (statutory employees, see page C-5). If you checked 32b, you MUST attach **Form 6198.**	32a ☐ All investment is at risk. 32b ☐ Some investment is not at risk.	

For Paperwork Reduction Act Notice, see Form 1040 instructions. Cat. No. 11334P **Schedule C (Form 1040) 1992**

A-4a Form 1040: Schedule C Profit or Loss from Business (front).

Department of the Treasury—Internal Revenue Service

Schedule C (Form 1040) 1992 Page **2**

Part III **Cost of Goods Sold** (see page C-5)

Line	Description		Amount
33	Inventory at beginning of year. If different from last year's closing inventory, attach explanation	33	
34	Purchases less cost of items withdrawn for personal use	34	
35	Cost of labor. Do not include salary paid to yourself	35	
36	Materials and supplies	36	
37	Other costs	37	
38	Add lines 33 through 37	38	
39	Inventory at end of year	39	
40	**Cost of goods sold.** Subtract line 39 from line 38. Enter the result here and on page 1, line 4	40	

Part IV **Principal Business or Professional Activity Codes**

Locate the major category that best describes your activity. Within the major category, select the activity code that most closely identifies the business or profession that is the principal source of your sales or receipts. **Enter this 4-digit code on page 1, line B.** For example, real estate agent is under the major category of **"Real Estate,"** and the code is "5520." **Note:** *If your principal source of income is from farming activities, you should file* ***Schedule F*** *(Form 1040), Profit or Loss From Farming.*

Agricultural Services, Forestry, Fishing

Code

1990 Animal services, other than breeding
1933 Crop services
2113 Farm labor & management services
2246 Fishing, commercial
2238 Forestry, except logging
2212 Horticulture & landscaping
2469 Hunting & trapping
1974 Livestock breeding
0836 Logging
1958 Veterinary services, including pets

Construction

0018 Operative builders (for own account)

Building Trade Contractors, Including Repairs

0414 Carpentering & flooring
0455 Concrete work
0273 Electrical work
0299 Masonry, dry wall, stone, & tile
0257 Painting & paper hanging
0232 Plumbing, heating, & air conditioning
0430 Roofing, siding & sheet metal
0885 Other building trade contractors (excavation, glazing, etc.)

General Contractors

0075 Highway & street construction
0059 Nonresidential building
0034 Residential building
3889 Other heavy construction (pipe laying, bridge construction, etc.)

Finance, Insurance, & Related Services

6064 Brokers & dealers of securities
6080 Commodity contracts brokers & dealers; security & commodity exchanges
6148 Credit institutions & mortgage bankers
5702 Insurance agents or brokers
5744 Insurance services (appraisal, consulting, inspection, etc.)
6130 Investment advisors & services
5777 Other financial services

Manufacturing, Including Printing & Publishing

0679 Apparel & other textile products
1115 Electric & electronic equipment
1073 Fabricated metal products
0638 Food products & beverages
0810 Furniture & fixtures
0695 Leather footwear, handbags, etc.
0836 Lumber & other wood products
1099 Machinery & machine shops
0877 Paper & allied products
1057 Primary metal industries
0851 Printing & publishing
1032 Stone, clay, & glass products
0653 Textile mill products
1883 Other manufacturing industries

Mining & Mineral Extraction

1537 Coal mining
1511 Metal mining
1552 Oil & gas
1719 Quarrying & nonmetallic mining

Real Estate

5538 Operators & lessors of buildings, including residential
5553 Operators & lessors of other real property
5520 Real estate agents & brokers
5579 Real estate property managers
5710 Subdividers & developers, except cemeteries
6155 Title abstract offices

Services: Personal, Professional, & Business Services

Amusement & Recreational Services

9670 Bowling centers
9688 Motion picture & tape distribution & allied services
9597 Motion picture & video production
9639 Motion picture theaters
8557 Physical fitness facilities
9696 Professional sports & racing, including promoters & managers
9811 Theatrical performers, musicians, agents, producers & related services
9613 Video tape rental
9837 Other amusement & recreational services

Automotive Services

8813 Automotive rental or leasing, without driver
8953 Automotive repairs, general & specialized
8839 Parking, except valet
8896 Other automotive services (wash, towing, etc.)

Business & Personal Services

7658 Accounting & bookkeeping
7716 Advertising, except direct mail
7682 Architectural services
8318 Barber shop (or barber)
8110 Beauty shop (or beautician)
8714 Child day care
7872 Computer programming, processing, data preparation & related services
7922 Computer repair, maintenance, & leasing
7286 Consulting services
7799 Consumer credit reporting & collection services
8755 Counseling (except health practitioners)
7732 Employment agencies & personnel supply
7518 Engineering services
7773 Equipment rental & leasing (except computer or automotive)
8532 Funeral services & crematories
7633 Income tax preparation
7914 Investigative & protective services
7617 Legal services (or lawyer)
7856 Mailing, reproduction, commercial art, photography, & stenographic services
7245 Management services
8771 Ministers & chaplains
8334 Photographic studios
7260 Public relations
8733 Research services
7708 Surveying services
8730 Teaching or tutoring
7880 Other business services
6882 Other personal services

Hotels & Other Lodging Places

7237 Camps & camping parks
7096 Hotels, motels, & tourist homes
7211 Rooming & boarding houses

Laundry & Cleaning Services

7450 Carpet & upholstery cleaning
7419 Coin-operated laundries & dry cleaning
7435 Full-service laundry, dry cleaning, & garment service
7476 Janitorial & related services (building, house, & window cleaning)

Medical & Health Services

9274 Chiropractors
9233 Dentist's office or clinic
9217 Doctor's (M.D.) office or clinic
9456 Medical & dental laboratories
9472 Nursing & personal care facilities
9290 Optometrists
9258 Osteopathic physicians & surgeons
9241 Podiatrists
9415 Registered & practical nurses
9431 Offices & clinics of other health practitioners (dieticians, midwives, speech pathologists, etc.)
9886 Other health services

Miscellaneous Repair, Except Computers

9019 Audio equipment & TV repair
9035 Electrical & electronic equipment repair, except audio & TV
9050 Furniture repair & reupholstery
2881 Other equipment repair

Trade, Retail—Selling Goods to Individuals & Households

3038 Catalog or mail order
3012 Selling door to door, by telephone or party plan, or from mobile unit
3053 Vending machine selling

Selling From Showroom, Store, or Other Fixed Location

Apparel & Accessories

3921 Accessory & specialty stores & furriers for women
3939 Clothing, family
3772 Clothing, men's & boys'
3913 Clothing, women's
3756 Shoe stores
3954 Other apparel & accessory stores

Automotive & Service Stations

3558 Gasoline service stations
3319 New car dealers (franchised)
3533 Tires, accessories, & parts
3335 Used car dealers
3517 Other automotive dealers (motorcycles, recreational vehicles, etc.)

Building, Hardware, & Garden Supply

4416 Building materials dealers
4457 Hardware stores
4473 Nurseries & garden supply stores
4432 Paint, glass, & wallpaper stores

Food & Beverages

0612 Bakeries selling at retail
3086 Catering services
3095 Drinking places (bars, taverns, pubs, saloons, etc.)
3079 Eating places, meals & snacks
3210 Grocery stores (general line)
3251 Liquor stores
3236 Specialized food stores (meat, produce, candy, health food, etc.)

Furniture & General Merchandise

3988 Computer & software stores
3970 Furniture stores
4317 Home furnishings stores (china, floor coverings, drapes)
4119 Household appliance stores
4333 Music & record stores
3996 TV, audio & electronic stores
3715 Variety stores
3731 Other general merchandise stores

Miscellaneous Retail Stores

4812 Boat dealers
5017 Book stores, excluding newsstands
4853 Camera & photo supply stores
3277 Drug stores
5058 Fabric & needlework stores
4655 Florists
5090 Fuel dealers (except gasoline)
4630 Gift, novelty & souvenir shops
4838 Hobby, toy, & game shops
4671 Jewelry stores
4895 Luggage & leather goods stores
5074 Mobile home dealers
4879 Optical goods stores
4697 Sporting goods & bicycle shops
5033 Stationery stores
4614 Used merchandise & antique stores (except motor vehicle parts)
5884 Other retail stores

Trade, Wholesale—Selling Goods to Other Businesses, etc.

Durable Goods, Including Machinery Equipment, Wood, Metals, etc.

2634 Agent or broker for other firms—more than 50% of gross sales on commission
2618 Selling for your own account

Nondurable Goods, Including Food, Fiber, Chemicals, etc.

2675 Agent or broker for other firms—more than 50% of gross sales on commission
2659 Selling for your own account

Transportation, Communications, Public Utilities, & Related Services

6619 Air transportation
6312 Bus & limousine transportation
6676 Communication services
6395 Courier or package delivery
6361 Highway passenger transportation (except chartered service)
6536 Public warehousing
6114 Taxicabs
6510 Trash collection without own dump
6635 Travel agents & tour operators
6338 Trucking (except trash collection)
6692 Utilities (dumps, snow plowing, road cleaning, etc.)
6551 Water transportation
6650 Other transportation services

8888 **Unable to classify**

☆U.S. GPO: 1992-43-1410168/315-174

A-4b Form 1040: Schedule C Profit or Loss from Business (back).

Department of the Treasury—Internal Revenue Service

SCHEDULES A&B (Form 1040)

Department of the Treasury Internal Revenue Service (B)

Schedule A—Itemized Deductions

(Schedule B is on back)

▶ Attach to Form 1040. ▶ See Instructions for Schedules A and B (Form 1040).

OMB No. 1545-0074

1992

Attachment Sequence No. 07

Name(s) shown on Form 1040 | Your social security number

Medical and Dental Expenses		**Caution:** *Do not include expenses reimbursed or paid by others.*		
	1	Medical and dental expenses (see page A-1)	1	
	2	Enter amount from Form 1040, line 32. 2		
	3	Multiply line 2 above by 7.5% (.075)	3	
	4	Subtract line 3 from line 1. If zero or less, enter -0- ▶		4
Taxes You Paid (See page A-1.)	5	State and local income taxes	5	
	6	Real estate taxes (see page A-2)	6	
	7	Other taxes. List—include personal property taxes ▶	7	
	8	Add lines 5 through 7 ▶		8
Interest You Paid (See page A-2.)	9a	Home mortgage interest and points reported to you on Form 1098	9a	
	b	Home mortgage interest not reported to you on Form 1098. If paid to an individual, show that person's name and address. ▶	9b	
Note: Personal interest is not deductible.	10	Points not reported to you on Form 1098. See page A-3 for special rules	10	
	11	Investment interest. If required, attach Form 4952. (See page A-3.)	11	
	12	Add lines 9a through 11 ▶		12
Gifts to Charity (See page A-3.)		**Caution:** *If you made a charitable contribution and received a benefit in return, see page A-3.*		
	13	Contributions by cash or check	13	
	14	Other than by cash or check. If over $500, you **MUST** attach Form 8283	14	
	15	Carryover from prior year	15	
	16	Add lines 13 through 15 ▶		16
Casualty and Theft Losses	17	Casualty or theft loss(es). Attach Form 4684. (See page A-4.) ▶		17
Moving Expenses	18	Moving expenses. Attach Form 3903 or 3903F. (See page A-4.) ▶		18
Job Expenses and Most Other Miscellaneous Deductions (See page A-5 for expenses to deduct here.)	19	Unreimbursed employee expenses—job travel, union dues, job education, etc. If required, you **MUST** attach Form 2106. (See page A-4.) ▶	19	
	20	Other expenses—investment, tax preparation, safe deposit box, etc. List type and amount ▶	20	
	21	Add lines 19 and 20	21	
	22	Enter amount from Form 1040, line 32. 22		
	23	Multiply line 22 above by 2% (.02)	23	
	24	Subtract line 23 from line 21. If zero or less, enter -0- ▶		24
Other Miscellaneous Deductions	25	Other—from list on page A-5. List type and amount ▶		25
Total Itemized Deductions	26	Is the amount on Form 1040, line 32, more than $105,250 (more than $52,625 if married filing separately)? • **NO.** Your deduction is not limited. Add lines 4, 8, 12, 16, 17, 18, 24, and 25. • **YES.** Your deduction may be limited. See page A-5 for the amount to enter. ▶		26
		Caution: *Be sure to enter on Form 1040, line 34, the* ***LARGER*** *of the amount on line 26 above or your standard deduction.*		

For Paperwork Reduction Act Notice, see Form 1040 instructions. Cat. No. 12611D **Schedule A (Form 1040) 1992**

A-5a Form 1040: Schedules A&B Schedule A—Itemized Deductions Schedule A (front).

Department of the Treasury—Internal Revenue Service

Schedules A&B (Form 1040) 1992 OMB No. 1545-0074 Page **2**

Name(s) shown on Form 1040. Do not enter name and social security number if shown on other side. | **Your social security number**

Schedule B—Interest and Dividend Income

Attachment Sequence No. **08**

Part I Interest Income

(See pages 14 and B-1.)

Note: If you received a Form 1099-INT, Form 1099-OID, or substitute statement from a brokerage firm, list the firm's name as the payer and enter the total interest shown on that form.

If you had over $400 in taxable interest income OR are claiming the exclusion of interest from series EE U.S. savings bonds issued after 1989, you must complete this part. List ALL interest you received. If you had over $400 in taxable interest income, you must also complete Part III. If you received, as a nominee, interest that actually belongs to another person, or you received or paid accrued interest on securities transferred between interest payment dates, see page B-1.

	Interest Income		Amount
1	List name of payer—if any interest income is from seller-financed mortgages, see page B-1 and list this interest first ▶	1	
2	Add the amounts on line 1	2	
3	Excludable interest on series EE U.S. savings bonds issued after 1989 from Form 8815, line 14. You MUST attach Form 8815 to Form 1040	3	
4	Subtract line 3 from line 2. Enter the result here and on Form 1040, line 8a ▶	4	

Part II Dividend Income

(See pages 15 and B-1.)

Note: If you received a Form 1099-DIV or substitute statement from a brokerage firm, list the firm's name as the payer and enter the total dividends shown on that form.

If you had over $400 in gross dividends and/or other distributions on stock, you must complete this part and Part III. If you received, as a nominee, dividends that actually belong to another person, see page B-1.

	Dividend Income		Amount
5	List name of payer—include on this line capital gain distributions, nontaxable distributions, etc. ▶	5	
6	Add the amounts on line 5	6	
7	Capital gain distributions. Enter here and on Schedule D* (7)		
8	Nontaxable distributions. (See the inst. for Form 1040, line 9.) (8)		
9	Add lines 7 and 8	9	
10	Subtract line 9 from line 6. Enter the result here and on Form 1040, line 9 ▶	10	

**If you received capital gain distributions but do not need Schedule D to report any other gains or losses, see the instructions for Form 1040, lines 13 and 14.*

Part III Foreign Accounts and Foreign Trusts

(See page B-2.)

If you had over $400 of interest or dividends OR had a foreign account or were a grantor of, or a transferor to, a foreign trust, you must complete this part.

		Yes	No
11a	At any time during 1992, did you have an interest in or a signature or other authority over a financial account in a foreign country, such as a bank account, securities account, or other financial account? See page B-2 for exceptions and filing requirements for Form TD F 90-22.1		
b	If "Yes," enter the name of the foreign country ▶		
12	Were you the grantor of, or transferor to, a foreign trust that existed during 1992, whether or not you have any beneficial interest in it? If "Yes," you may have to file Form 3520, 3520-A, or 926		

For Paperwork Reduction Act Notice, see Form 1040 instructions. **Schedule B (Form 1040) 1992**

*U.S. Government Printing Office: 1992 — 315-032

A-5b Form 1040: Schedules A&B Schedule B—Interest and Dividend Income (back).

Department of the Treasury—Internal Revenue Service

SCHEDULE E
(Form 1040)

Department of the Treasury
Internal Revenue Service (0)

Supplemental Income and Loss

(From rental real estate, royalties, partnerships, estates, trusts, REMICs, etc.)

▶ Attach to Form 1040 or Form 1041.
▶ See Instructions for Schedule E (Form 1040).

OMB No. 1545-0074

1992

Attachment Sequence No. **13**

Name(s) shown on return | **Your social security number**

Part I **Income or Loss From Rental Real Estate and Royalties** **Note:** *Report income and expenses from the rental of personal property on* ***Schedule C*** *or* ***C-EZ.*** *Report farm rental income or loss from* ***Form 4835*** *on page 2, line 39.*

1	Show the kind and location of each **rental real estate property:**	2	For each rental real estate property listed on line 1, did you or your family use it for personal purposes for more than the greater of 14 days or 10% of the total days rented at fair rental value during the tax year? (See page E-1.)		Yes	No
A				A		
B				B		
C				C		

Income:		Properties A	Properties B	Properties C	Totals (Add columns A, B, and C.)	
3 Rents received	3				3	
4 Royalties received	4				4	
Expenses:						
5 Advertising	5					
6 Auto and travel (see page E-2)	6					
7 Cleaning and maintenance	7					
8 Commissions	8					
9 Insurance	9					
10 Legal and other professional fees	10					
11 Management fees	11					
12 Mortgage interest paid to banks, etc. (see page E-2)	12				12	
13 Other interest	13					
14 Repairs	14					
15 Supplies	15					
16 Taxes	16					
17 Utilities	17					
18 Other (list) ▶	18					
19 Add lines 5 through 18	19				19	
20 Depreciation expense or depletion (see page E-2)	20				20	
21 Total expenses. Add lines 19 and 20	21					
22 Income or (loss) from rental real estate or royalty properties. Subtract line 21 from line 3 (rents) or line 4 (royalties). If the result is a (loss), see page E-2 to find out if you must file **Form 6198**	22					
23 Deductible rental real estate loss. **Caution:** *Your rental real estate loss on line 22 may be limited. See page E-3 to find out if you must file* ***Form 8582***	23	( )	( )	( )		
24 **Income.** Add positive amounts shown on line 22. **Do not** include any losses					24	
25 **Losses.** Add royalty losses from line 22 and rental real estate losses from line 23. Enter the total losses here					25	( )
26 Total rental real estate and royalty income or (loss). Combine lines 24 and 25. Enter the result here. If Parts II, III, IV, and line 39 on page 2 do not apply to you, also enter this amount on Form 1040, line 18. Otherwise, include this amount in the total on line 40 on page 2					26	

For Paperwork Reduction Act Notice, see Form 1040 instructions. Cat. No. 11344L **Schedule E (Form 1040) 1992**

A-6a Form 1040: Schedule E Supplemental Income and Loss (front).

Department of the Treasury—Internal Revenue Service

Name(s) shown on return. Do not enter name and social security number if shown on other side. | **Your social security number**

Note: *If you report amounts from farming or fishing on Schedule E, you must enter your gross income from those activities on line 41 below.*

Part II Income or Loss From Partnerships and S Corporations

If you report a loss from an at-risk activity, you MUST check either column **(e)** or **(f)** of line 27 to describe your investment in the activity. See page E-3. If you check column **(f)**, you must attach **Form 6198**.

27	(a) Name	(b) Enter P for partnership; S for S corporation	(c) Check if foreign partnership	(d) Employer identification number	Investment At Risk? (e) All is at risk	(f) Some is not at risk
A						
B						
C						
D						
E						

	Passive Income and Loss		Nonpassive Income and Loss		
	(g) Passive loss allowed (attach Form 8582 if required)	(h) Passive income from Schedule K-1	(i) Nonpassive loss from Schedule K-1	(j) Section 179 expense deduction from Form 4562	(k) Nonpassive income from Schedule K-1
A					
B					
C					
D					
E					
28a Totals					
b Totals					

29	Add columns (h) and (k) of line 28a	29
30	Add columns (g), (i), and (j) of line 28b	30 ( )
31	Total partnership and S corporation income or (loss). Combine lines 29 and 30. Enter the result here and include in the total on line 40 below	31

Part III Income or Loss From Estates and Trusts

32	(a) Name	(b) Employer identification number
A		
B		
C		

	Passive Income and Loss		Nonpassive Income and Loss	
	(c) Passive deduction or loss allowed (attach Form 8582 if required)	(d) Passive income from Schedule K-1	(e) Deduction or loss from Schedule K-1	(f) Other income from Schedule K-1
A				
B				
C				
33a Totals				
b Totals				

34	Add columns (d) and (f) of line 33a	34
35	Add columns (c) and (e) of line 33b	35 ( )
36	Total estate and trust income or (loss). Combine lines 34 and 35. Enter the result here and include in the total on line 40 below	36

Part IV Income or Loss From Real Estate Mortgage Investment Conduits (REMICs)—Residual Holder

37	(a) Name	(b) Employer identification number	(c) Excess inclusion from Schedules Q, line 2c (see page E-4)	(d) Taxable income (net loss) from Schedules Q, line 1b	(e) Income from Schedules Q, line 3b

38	Combine columns (d) and (e) only. Enter the result here and include in the total on line 40 below	38

Part V Summary

39	Net farm rental income or (loss) from **Form 4835**. Also, complete line 41 below	39
40	TOTAL income or (loss). Combine lines 26, 31, 36, 38, and 39. Enter the result here and on Form 1040, line 18 ▶	40
41	**Reconciliation of Farming and Fishing Income:** Enter your **gross** farming and fishing income reported in Parts II and III and on line 39 (see page E-4)	41

*U.S. Government Printing Office: 1992 — 315-180

A-6b Form 1040: Schedule E Supplemental Income and Loss (back).

Department of the Treasury—Internal Revenue Service

Form **2106** — Department of the Treasury, Internal Revenue Service (0)

Employee Business Expenses

▶ See separate instructions.

▶ Attach to Form 1040.

OMB No. 1545-0139 — **1991** — Attachment Sequence No. **54**

Your name	Social security number	Occupation in which expenses were incurred

Part I **Employee Business Expenses and Reimbursements**

STEP 1 Enter Your Expenses		**Column A** Other Than Meals and Entertainment	**Column B** Meals and Entertainment
1 Vehicle expense from line 22 or line 29	1		
2 Parking fees, tolls, and local transportation, including train, bus, etc.	2		
3 Travel expense while away from home overnight, including lodging, airplane, car rental, etc. **Do not** include meals and entertainment	3		
4 Business expenses not included on lines 1 through 3. **Do not** include meals and entertainment	4		
5 Meals and entertainment expenses. (See instructions.)	5		
6 **Total expenses.** In Column A, add lines 1 through 4 and enter the result. In Column B, enter the amount from line 5	6		

Note: *If you were not reimbursed for any expenses in Step 1, skip line 7 and enter the amount from line 6 on line 8.*

STEP 2 Enter Amounts Your Employer Gave You for Expenses Listed in STEP 1			
7 Enter amounts your employer gave you that were **not** reported to you in Box 10 of Form W-2. Include any amount reported under code "L" in Box 17 of your Form W-2. (See instructions.)	7		

STEP 3 Figure Expenses To Deduct on Schedule A (Form 1040)			
8 Subtract line 7 from line 6	8		
Note: *If **both columns** of line 8 are zero, **stop here.** If Column A is less than zero, report the amount as income and enter -0- on line 10, Column A. See the instructions for how to report.*			
9 Enter 20% (.20) of line 8, Column B	9		
10 Subtract line 9 from line 8	10		
11 Add the amounts on line 10 of both columns and enter the total here. **Also enter the total on Schedule A (Form 1040), line 19.** (Qualified performing artists and individuals with disabilities, see the instructions for special rules on where to enter the total.) ▶	11		

For Paperwork Reduction Act Notice, see instructions. Cat. No. 11700N Form **2106** (1991)

A-7a Form 2106: Employee Business Expenses (front).

Department of the Treasury—Internal Revenue Service

Part II **Vehicle Expenses (See instructions to find out which sections to complete.)**

Section A.—General Information		**(a) Vehicle 1**	**(b) Vehicle 2**
12 Enter the date vehicle was placed in service	12	/ /	/ /
13 Total mileage vehicle was used during 1991	13	miles	miles
14 Miles included on line 13 that vehicle was used for business	14	miles	miles
15 Percent of business use (divide line 14 by line 13)	15	%	%
16 Average daily round trip commuting distance	16	miles	miles
17 Miles included on line 13 that vehicle was used for commuting	17	miles	miles
18 Other personal mileage (add lines 14 and 17 and subtract the total from line 13)	18	miles	miles

19 Do you (or your spouse) have another vehicle available for personal purposes? ☐ Yes ☐ No

20 If your employer provided you with a vehicle, is personal use during off duty hours permitted? ☐ Yes ☐ No ☐ Not applicable

21a Do you have evidence to support your deduction? ☐ Yes ☐ No **21b** If "Yes," is the evidence written? ☐ Yes ☐ No

Section B.—Standard Mileage Rate (Use this section only if you own the vehicle.)

22 Multiply line 14 by 27.5¢ (.275). Enter the result here and on line 1. (Rural mail carriers, see instructions.)	22	

Section C.—Actual Expenses		**(a) Vehicle 1**		**(b) Vehicle 2**	
23 Gasoline, oil, repairs, vehicle insurance, etc.	23				
24a Vehicle rentals	24a				
b Inclusion amount	24b				
c Subtract line 24b from line 24a	24c				
25 Value of employer-provided vehicle (applies only if 100% of annual lease value was included on Form W-2. See instructions.)	25				
26 Add lines 23, 24c, and 25	26				
27 Multiply line 26 by the percentage on line 15	27				
28 Enter amount from line 38 below	28				
29 Add lines 27 and 28. Enter total here and on line 1	29				

Section D.—Depreciation of Vehicles (Use this section only if you own the vehicle.)

		(a) Vehicle 1		**(b) Vehicle 2**	
30 Enter cost or other basis. (See instructions.)	30				
31 Enter amount of section 179 deduction. (See instructions.)	31				
32 Multiply line 30 by line 15. (See instructions if you elected the section 179 deduction.)	32				
33 Enter depreciation method and percentage. (See instructions.)	33				
34 Multiply line 32 by the percentage on line 33. (See instructions.)	34				
35 Add lines 31 and 34	35				
36 Enter the limitation amount from the table in the line 36 instructions	36				
37 Multiply line 36 by the percentage on line 15	37				
38 Enter the **smaller** of line 35 or line 37. Also enter the amount on line 28 above	38				

*U.S. Government Printing Office: 1991 — 285-282

A-7b Form 2106—Employee Business Expenses (back).

Department of the Treasury—Internal Revenue Service.

Form **4684**

Department of the Treasury
Internal Revenue Service

Casualties and Thefts

► See separate instructions.
► Attach to your tax return.
► Use a separate Form 4684 for each different casualty or theft.

OMB No. 1545-0177
1991
Attachment Sequence No. **26**

Name(s) shown on tax return | **Identifying number**

Note: *Use Section A for casualties and thefts of personal use property and Section B for business and income-producing property.*

SECTION A.—Personal Use Property *(Casualties and thefts of property **not** used in a trade or business or for income-producing purposes)*

1 Description of properties (show kind, location, and date acquired for each):
Property **A** ....
Property **B** ....
Property **C** ....
Property **D** ....

		Properties (use a separate column for each property lost or damaged from one casualty or theft)			
		A	B	C	D
2 Cost or other basis of each property	2				
3 Insurance or other reimbursement (whether or not you submitted a claim). See instructions	3				
Note: *If line 2 is **more** than line 3, skip line 4.*					
4 Gain from casualty or theft. If line 3 is **more than** line 2, enter the difference here and skip lines 5 through 9 for that column. (If line 3 includes an amount that you did not receive, see instructions.)	4				
5 Fair market value **before** casualty or theft	5				
6 Fair market value **after** casualty or theft	6				
7 Subtract line 6 from line 5	7				
8 Enter the **smaller** of line 2 or line 7	8				
9 Subtract line 3 from line 8 (If zero or less, enter -0-.)	9				

10 Casualty or theft loss. Add the amounts on line 9. Enter the total	10	
11 Enter the amount from line 10 or $100, whichever is **smaller**	11	
12 Subtract line 11 from line 10	12	
Caution: *Use only one Form 4684 for lines 13 through 18.*		
13 Add the amounts on line 12 of all Forms 4684, Section A	13	
14 Combine the amounts from line 4 of all Forms 4684, Section A	14	
15 • If line 14 is **more than** line 13, enter the difference here and on Schedule D. Do not complete the rest of this section (see instructions). • If line 14 is **less than** line 13, enter -0- here and continue with the form. • If line 14 is **equal to** line 13, enter -0- here. Do not complete the rest of this section.	15	
16 If line 14 is **less than** line 13, enter the difference	16	
17 Enter 10% of your adjusted gross income (Form 1040, line 32). Estates and trusts, see instructions	17	
18 Subtract line 17 from line 16. If zero or less, enter -0-. Also enter result on Schedule A (Form 1040), line 17. Estates and trusts, enter on the "Other deductions" line of your tax return	18	

For Paperwork Reduction Act Notice, see page 1 of separate instructions. Cat. No. 12997O Form **4684** (1991)

A-8a Form 4684: Casualties and Thefts (front).

Department of the Treasury—Internal Revenue Service

Form 4684 (1991) Attachment Sequence No. **26** Page **2**

Name(s) shown on tax return. (Do not enter name and identifying number if shown on other side.) | **Identifying number**

SECTION B.—Business and Income-Producing Property *(Casualties and thefts of property used in a trade or business or for income-producing purposes)*

Part I Casualty or Theft Gain or Loss (Use a separate Part I for each casualty or theft)

1 Description of properties (show kind, location, and date acquired for each):

Property **A** ....

Property **B** ....

Property **C** ....

Property **D** ....

		Properties (use a separate column for each property lost or damaged from one casualty or theft)			
		A	**B**	**C**	**D**
2 Cost or adjusted basis of each property	2				
3 Insurance or other reimbursement (whether or not you submitted a claim). See the instructions for Section A, line 3	3				
Note: *If line 2 is* ***more*** *than line 3, skip line 4.*					
4 Gain from casualty or theft. If line 3 is **more than** line 2, enter the difference here and on line 11 or line 16, column **(c)**, except as provided in the instructions for line 15. Also, skip lines 5 through 9 for that column. (If line 3 includes an amount that you did not receive, see the instructions for Section A, line 4.)	4				
5 Fair market value **before** casualty or theft	5				
6 Fair market value **after** casualty or theft	6				
7 Subtract line 6 from line 5	7				
8 Enter the **smaller** of line 2 or line 7	8				
Note: *If the property was totally destroyed by casualty, or lost from theft, enter on line 8 the amount from line 2.*					
9 Subtract line 3 from line 8 (If zero or less, enter -0-.)	9				

10 Casualty or theft loss. Add the amounts on line 9. Enter the total here and on line 11 **or** line 16 (see instructions). | **10** |

Part II Summary of Gains and Losses (from separate Parts I)

	(a) Identify casualty or theft		**(b)** Losses from casualties or thefts			**(c)** Gains from casualties or thefts includible in income
			(i) Trade, business, rental or royalty property	*(ii)* Income-producing property		
	Casualty or Theft of Property Held One Year or Less					
11			( )	( )		
			( )	( )		
12	Totals. Add the amounts on line 11	12	( )	( )		
13	Combine line 12, columns (b)(i) and (c). Enter the net gain or (loss) here and on Form 4797, Part II, line 14. (If Form 4797 is not otherwise required, see instructions.)				13	
14	Enter the amount from line 12, column (b)(ii) here and on Schedule A (Form 1040), line 20. Partnerships, S corporations, estates and trusts, see instructions				14	
	Casualty or Theft of Property Held More Than One Year					
15	Casualty or theft gains from Form 4797, Part III, line 32				15	
16			( )	( )		
			( )	( )		
17	Total losses. Add amounts on line 16, columns (b)(i) and (b)(ii)	17	( )	( )		
18	Total gains. Add lines 15 and 16, column (c)				18	
19	Add amounts on line 17, columns (b)(i) and (b)(ii)				19	
20	If the loss on line 19 is **more than** the gain on line 18:					
a	Combine line 17, column (b)(i) and line 18, and enter the net gain or (loss) here. Partnerships and S corporations see the note below. All others enter this amount on Form 4797, Part II, line 14. (If Form 4797 is not otherwise required, see instructions.)				20a	
b	Enter the amount from line 17, column (b)(ii) here. Partnerships and S corporations see the note below. Individuals enter this amount on Schedule A (Form 1040), line 20. Estates and trusts, enter on the "Other deductions" line of your tax return				20b	
21	If the loss on line 19 is **equal to** or **less than** the gain on line 18, combine these lines and enter here. Partnerships see the note below. All others enter this amount on Form 4797, Part I, line 3				21	

Note: *Partnerships, enter the amount from line 20a, 20b, or line 21 on Form 1065, Schedule K, line 7. S corporations, enter the amount from line 20a or 20b on Form 1120S, Schedule K, line 6.*

*U.S. Government Printing Office: 1991 — 285-336

A-8b Form 4684: Casualties and Thefts (back).

Department of the Treasury—Internal Revenue Service

Form **4562**

Department of the Treasury
Internal Revenue Service (0)

Depreciation and Amortization
(Including Information on Listed Property)

► See separate instructions. ► Attach this form to your return.

OMB No. 1545-0172

1992

Attachment Sequence No. **67**

Name(s) shown on return

Identifying number

Business or activity to which this form relates

Part I **Election To Expense Certain Tangible Property (Section 179) (Note:** *If you have any "Listed Property," complete Part V before you complete Part I.)*

1	Maximum dollar limitation (see instructions)	1	$10,000
2	Total cost of section 179 property placed in service during the tax year (see instructions)	2	
3	Threshold cost of section 179 property before reduction in limitation	3	$200,000
4	Reduction in limitation. Subtract line 3 from line 2, but do not enter less than -0-	4	
5	Dollar limitation for tax year. Subtract line 4 from line 1, but do not enter less than -0-	5	

	(a) Description of property	(b) Cost	(c) Elected cost
6			
7	Listed property. Enter amount from line 26	7	

8	Total elected cost of section 179 property. Add amounts in column (c), lines 6 and 7	8	
9	Tentative deduction. Enter the smaller of line 5 or line 8	9	
10	Carryover of disallowed deduction from 1991 (see instructions)	10	
11	Taxable income limitation. Enter the smaller of taxable income or line 5 (see instructions)	11	
12	Section 179 expense deduction. Add lines 9 and 10, but do not enter more than line 11	12	
13	Carryover of disallowed deduction to 1993. Add lines 9 and 10, less line 12 ► 13		

Note: *Do not use Part II or Part III below for automobiles, certain other vehicles, cellular telephones, computers, or property used for entertainment, recreation, or amusement (listed property). Instead, use Part V for listed property.*

Part II **MACRS Depreciation For Assets Placed in Service ONLY During Your 1992 Tax Year (Do Not Include Listed Property)**

(a) Classification of property	(b) Month and year placed in service	(c) Basis for depreciation (business/investment use only—see instructions)	(d) Recovery period	(e) Convention	(f) Method	(g) Depreciation deduction
14 General Depreciation System (GDS) (see instructions):						
a 3-year property						
b 5-year property						
c 7-year property						
d 10-year property						
e 15-year property						
f 20-year property						
g Residential rental property			27.5 yrs.	MM	S/L	
			27.5 yrs.	MM	S/L	
h Nonresidential real property			31.5 yrs.	MM	S/L	
			31.5 yrs.	MM	S/L	
15 Alternative Depreciation System (ADS) (see instructions):						
a Class life					S/L	
b 12-year			12 yrs.		S/L	
c 40-year			40 yrs.	MM	S/L	

Part III **Other Depreciation (Do Not Include Listed Property)**

16	GDS and ADS deductions for assets placed in service in tax years beginning before 1992 (see instructions)	16	
17	Property subject to section 168(f)(1) election (see instructions)	17	
18	ACRS and other depreciation (see instructions)	18	

Part IV **Summary**

19	Listed property. Enter amount from line 25	19	
20	**Total.** Add deductions on line 12, lines 14 and 15 in column (g), and lines 16 through 19. Enter here and on the appropriate lines of your return. (Partnerships and S corporations—see instructions)	20	
21	For assets shown above and placed in service during the current year, enter the portion of the basis attributable to section 263A costs (see instructions) 21		

For Paperwork Reduction Act Notice, see page 1 of the separate instructions. Cat. No. 12906N Form **4562** (1992)

A-9a Form 4562: Depreciation and Amortization (front).

Department of the Treasury—Internal Revenue Service

Form 4562 (1992) Page 2

Part V **Listed Property—Automobiles, Certain Other Vehicles, Cellular Telephones, Computers, and Property Used for Entertainment, Recreation, or Amusement**

*For any vehicle for which you are using the standard mileage rate or deducting lease expense, complete **only** 22a, 22b, columns (a) through (c) of Section A, all of Section B, and Section C if applicable.*

Section A—Depreciation (Caution: *See instructions for limitations for automobiles.)*

22a Do you have evidence to support the business/investment use claimed? ☐ **Yes** ☐ **No** **22b** If "Yes," is the evidence written? ☐ **Yes** ☐ **No**

(a) Type of property (list vehicles first)	(b) Date placed in service	(c) Business/ investment use percentage	(d) Cost or other basis	(e) Basis for depreciation (business/investment use only)	(f) Recovery period	(g) Method/ Convention	(h) Depreciation deduction	(i) Elected section 179 cost
23 Property used more than 50% in a qualified business use (see instructions):								
		%						
		%						
		%						
24 Property used 50% or less in a qualified business use (see instructions):								
		%				S/L –		
		%				S/L –		
		%				S/L –		

25 Add amounts in column (h). Enter the total here and on line 19, page 1 . . . . . . **25**

26 Add amounts in column (i). Enter the total here and on line 7, page 1 . . . . **26**

Section B—Information Regarding Use of Vehicles—*If you deduct expenses for vehicles:*

- *Always complete this section for vehicles used by a sole proprietor, partner, or other "more than 5% owner," or related person.*
- *If you provided vehicles to your employees, first answer the questions in Section C to see if you meet an exception to completing this section for those vehicles.*

	(a) Vehicle 1		(b) Vehicle 2		(c) Vehicle 3		(d) Vehicle 4		(e) Vehicle 5		(f) Vehicle 6	
27 Total business/investment miles driven during the year (DO NOT include commuting miles)												
28 Total commuting miles driven during the year												
29 Total other personal (noncommuting) miles driven												
30 Total miles driven during the year. Add lines 27 through 29.												
	Yes	**No**	**Yes**	**No**	**Yes**	**No**	**Yes**	**No**	**Yes**	**No**	**Yes**	**No**
31 Was the vehicle available for personal use during off-duty hours?												
32 Was the vehicle used primarily by a more than 5% owner or related person?												
33 Is another vehicle available for personal use?												

Section C—Questions for Employers Who Provide Vehicles for Use by Their Employees

Answer these questions to determine if you meet an exception to completing Section B. **Note:** *Section B must always be completed for vehicles used by sole proprietors, partners, or other more than 5% owners or related persons.*

	Yes	No
34 Do you maintain a written policy statement that prohibits all personal use of vehicles, including commuting, by your employees?		
35 Do you maintain a written policy statement that prohibits personal use of vehicles, except commuting, by your employees? (See instructions for vehicles used by corporate officers, directors, or 1% or more owners.)		
36 Do you treat all use of vehicles by employees as personal use?		
37 Do you provide more than five vehicles to your employees and retain the information received from your employees concerning the use of the vehicles?		
38 Do you meet the requirements concerning qualified automobile demonstration use (see instructions)? . .		

Note: *If your answer to 34, 35, 36, 37, or 38 is "Yes," you need not complete Section B for the covered vehicles.*

Part VI **Amortization**

(a) Description of costs	(b) Date amortization begins	(c) Amortizable amount	(d) Code section	(e) Amortization period or percentage	(f) Amortization for this year
39 Amortization of costs that begins during your 1992 tax year:					
40 Amortization of costs that began before 1992				**40**	
41 Total. Enter here and on "Other Deductions" or "Other Expenses" line of your return . . .				**41**	

☆ U.S. GOVERNMENT PRINTING OFFICE: 1992 315-330

A-9b Form 4562: Depreciation and Amortization (back).

Department of the Treasury—Internal Revenue Service

Form **8829**

Department of the Treasury
Internal Revenue Service

Expenses for Business Use of Your Home

▶ File with Schedule C (Form 1040).

▶ See instructions on back.

OMB No. 1545-1256

1991

Attachment Sequence No. **66**

Name of proprietor | Your social security number

Part I Part of Your Home Used for Business

1	Area used exclusively for business (see instructions). Include area used for inventory storage or as a day-care facility that does not meet exclusive use test		1	
2	Total area of home		2	
3	Divide line 1 by line 2. Enter the result as a percentage		3	%
	• For day-care facilities not used exclusively for business, also complete lines 4–6.			
	• All others, skip lines 4–6 and enter the amount from line 3 on line 7.			
4	Total hours facility used for day care during the year. Multiply days used by number of hours used per day	4 hr.		
5	Total hours available for use during the year (365 days x 24 hours) (see instructions).	5 8,760 hr.		
6	Divide line 4 by line 5. Enter the result as a decimal amount	6 .		
7	Business percentage. For day-care facilities not used exclusively for business, multiply line 6 by line 3 (enter the result as a percentage). All others, enter the amount from line 3 ▶		7	%

Part II Figure Your Allowable Deduction

Line	Description		(a) Direct expenses	(b) Indirect expenses		
8	Enter the amount from Schedule C, line 29. (If more than one place of business, see instructions.)				8	
9	Casualty losses	9				
10	Deductible mortgage interest	10				
11	Real estate taxes	11				
12	Add lines 9, 10, and 11	12				
13	Multiply line 12, column (b) by line 7			13		
14	Add line 12, column (a) and line 13				14	
15	Subtract line 14 from line 8. If zero or less, enter -0-				15	
16	Excess mortgage interest (see instructions)	16				
17	Insurance	17				
18	Repairs and maintenance	18				
19	Utilities	19				
20	Other expenses	20				
21	Add lines 16 through 20	21				
22	Multiply line 21, column (b) by line 7			22		
23	Carryover of operating expenses from 1990			23		
24	Add line 21 in column (a), line 22, and line 23				24	
25	Allowable operating expenses. Enter the **smaller** of line 15 or line 24				25	
26	Limit on excess casualty losses and depreciation. Subtract line 25 from line 15				26	
27	Excess casualty losses (see instructions)			27		
28	Depreciation of your home from Part III below			28		
29	Carryover of excess casualty losses and depreciation from 1990			29		
30	Add lines 27 through 29				30	
31	Allowable excess casualty losses and depreciation. Enter the **smaller** of line 26 or line 30				31	
32	Add lines 14, 25, and 31				32	
33	Casualty losses included on lines 14 and 31. (Carry this amount to **Form 4684,** Section B.)				33	
34	Allowable expenses for business use of your home. Subtract line 33 from line 32. Enter here and on Schedule C, line 30 ▶				34	

Part III Depreciation of Your Home

35	Enter the **smaller** of your home's adjusted basis or its fair market value (see instructions)	35	
36	Value of land included on line 35	36	
37	Basis of building. Subtract line 36 from line 35	37	
38	Business basis of building. Multiply line 37 by line 7	38	
39	Depreciation percentage (see instructions)	39	%
40	Depreciation allowable. Multiply line 38 by the percentage on line 39. Enter here and on line 28 above	40	

Part IV Carryover of Unallowed Expenses to 1992

41	Operating expenses. Subtract line 25 from line 24. If less than zero, enter -0-	41	
42	Excess casualty losses and depreciation. Subtract line 31 from line 30. If less than zero, enter -0-	42	

For Paperwork Reduction Act Notice, see back of form. Cat. No. 13232M Form **8829** (1991)

A-10a Form 8829: Expenses for Business Use of Your Home (front).

Department of the Treasury—Internal Revenue Service

General Instructions

Paperwork Reduction Act Notice.—We ask for the information on this form to carry out the Internal Revenue laws of the United States. You are required to give us the information. We need it to ensure that you are complying with these laws and to allow us to figure and collect the right amount of tax.

The time needed to complete and file this form will vary depending on individual circumstances. The estimated average time is: **Recordkeeping,** 52 min.; **Learning about the law or the form,** 7 min.; **Preparing the form,** 1 hr., 13 min.; and **Copying, assembling, and sending the form to the IRS,** 20 min.

If you have comments concerning the accuracy of these time estimates or suggestions for making this form more simple, we would be happy to hear from you. You can write to both the IRS and the Office of Management and Budget at the addresses listed in the instructions for Form 1040.

Purpose of Form

Use Form 8829 to figure the allowable expenses for business use of your home on **Schedule C** (Form 1040) and any carryover to 1992 of amounts not deductible in 1991.

You must meet specific requirements to deduct expenses for the business use of your home. Even if you meet these requirements, your deductible expenses are limited. For details, get **Pub. 587,** Business Use of Your Home.

Who May Deduct Expenses for Business Use of a Home

General rule.—You may deduct business expenses that apply to a part of your home **only** if that part is exclusively used on a regular basis:

- As your principal place of business for any of your trades or businesses; or
- As a place of business used by your patients, clients, or customers to meet or deal with you in the normal course of your trade or business; or
- In connection with your trade or business if it is a separate structure that is not attached to your home.

Exception for storage of inventory.—You may also deduct expenses that apply to space within your home if it is the **only** fixed location of your trade or business. The space must be used on a regular basis to store inventory from your trade or business of selling products at retail or wholesale.

Exception for day-care facilities.—If you use space in your home on a regular basis in your trade or business of providing day care, you may be able to deduct the business expenses even though you use the same space for nonbusiness purposes.

Specific Instructions

Part I

Lines 1 and 2.—You may use square feet to determine the area on lines 1 and 2. If the rooms in your home are about the same size, you may figure area using the number of rooms instead of square feet. You may use any other reasonable method if it accurately figures your business percentage on line 7.

Line 4.—Enter the total number of hours the facility was used for day care during the year.

Example. Your home is used Monday through Friday for 12 hours per day for 250 days during the year. It is also used on 50 Saturdays for 8 hours per day. Enter 3,400 hours on line 4 (3,000 hours for weekdays plus 400 hours for Saturdays).

Line 5.—If you started or stopped using your home for day care in 1991, you must prorate the number of hours based on the number of days the home was available for day care. Cross out the preprinted entry on line 5. Multiply 24 hours by the number of days available and enter the result.

Part II

Enter as direct or indirect expenses only expenses for the business use of your home (i.e., expenses allowable only because your home is used for business). Other expenses, such as salaries, supplies, and business telephone expenses, which are deductible elsewhere on Schedule C, should not be entered on Form 8829.

Direct expenses benefit only the business part of your home. They include painting or repairs made to the specific area or room used for business. Enter 100% of your direct expenses on the appropriate expense line in column (a).

Indirect expenses are for keeping up and running your entire home. They benefit both the business and personal parts of your home. Generally, enter 100% of your indirect expenses on the appropriate expense line in column (b). **Exception:** If the business percentage of an indirect expense is different from the percentage on line 7, enter only the business part of the expense on the appropriate line in column (a), and leave that line in column (b) blank. For example, your electric bill is $800 for lighting, cooking, laundry, and television. If you reasonably estimate $300 of your electric bill is for lighting and you use 10% of your home for business, enter $30 on line 19 in column (a) and leave line 19 in column (b) blank.

Line 8.—If all of the gross income from your trade or business is from the business use of your home, enter on line 8 the amount from Schedule C, line 29.

If part of the income is from a place of business other than your home, you must first determine the part of your gross income (Schedule C, line 7) from the business use of your home. In making this determination, consider the amount of time you spend at each location as well as other facts. After determining the part of your gross income from the business use of your home, subtract from that amount the total from Schedule C, line 28. Enter the result on line 8 of Form 8829.

Lines 9, 10, and 11.—Enter only the amounts that would be deductible whether or not you used your home for business (i.e., amounts allowable as itemized deductions on **Schedule A** (Form 1040)).

Treat **casualty losses** as personal expenses for this step. Figure the amount to enter on line 9 by completing Form 4684, Section A. When figuring line 17 of Section A, enter 10% of your adjusted gross income excluding the gross income from business use of your home and the deductions attributable to that income. Include on line 9 of Form 8829 the amount from Form 4684, Section A, line 18. See line 27 to deduct part of the casualty losses not allowed because of the limits on Form 4684, Section A.

Do not file or use that Form 4684 to figure the amount of casualty losses to deduct on Schedule A. Instead, complete a separate Form 4684 to deduct the personal portion of your casualty losses.

On line 10, include only **mortgage interest** that would be deductible on Schedule A and that qualifies as a direct or indirect expense. Do not include interest on a mortgage loan that did not benefit your home (e.g., a home equity loan used to pay off credit card bills, to buy a car, or to pay tuition costs).

Line 16.—If the amount of home mortgage interest you deduct on Schedule A is limited, enter the part of the excess mortgage interest that qualifies as a direct or indirect expense. Do not include mortgage interest on a loan that did not benefit your home (explained above).

Line 20.—If you rent rather than own your home, include the rent you paid on line 20, column (b).

Line 23.—If you were unable to deduct all of your 1990 operating expenses due to the limit on the deductible amount, enter on line 23 the amount of operating expenses you are carrying forward to 1991.

Line 27.—Multiply your casualty losses in excess of the amount on line 9 by the business percentage of those losses and enter the result.

Line 29.—If you were unable to deduct all of your 1990 excess casualty losses and depreciation due to the limit on the deductible amount, enter the amount of excess depreciation and casualty losses you are carrying forward to 1991.

Part III

Lines 35 through 37.—Enter on line 35 the cost or other basis of your home, or if less, the fair market value of your home on the date you first used the home for business. **Do not** adjust this amount for depreciation claimed or changes in fair market value after the year you first used your home for business. Allocate this amount between land and building values on lines 36 and 37.

Show on an attached schedule the cost or other basis of additions and improvements placed in service after you began to use your home for business. Do not include any amounts on lines 35 through 38 for these expenditures. Instead, see the instructions for line 40.

Line 39.—If you first used your home for business in 1991, enter the percentage for the month you first used it for business.

Jan.	3.042%	**July**	1.455%
Feb.	2.778%	**Aug.**	1.190%
March	2.513%	**Sept.**	0.926%
April	2.249%	**Oct.**	0.661%
May	1.984%	**Nov.**	0.397%
June	1.720%	**Dec.**	0.132%

If you first used your home for business before 1991 and after 1986, enter 3.175%. If the business use began before 1987 or you stopped using your home for business before the end of the year, see **Pub. 534,** Depreciation, for the percentage to enter.

Line 40.—Include on line 40 depreciation on additions and improvements placed in service after you began using your home for business. See Pub. 534 to figure the amount of depreciation allowed on these expenditures. Attach a schedule showing how you figured depreciation on any additions or improvements. Write "See attached" below the entry space.

Complete and attach **Form 4562,** Depreciation and Amortization, if you first used your home for business in 1991 or you are depreciating additions or improvements placed in service in 1991. If you first used your home for business in 1991, enter on Form 4562, in column (c) of line 14h, the amount from line 38 of Form 8829. Then enter on Form 4562, in column (g) of line 14h, the amount from line 40 of Form 8829.

A-10b Form 8829: Expenses for Business Use of Your Home (back).

Department of thTreasury—Internal Revenue Service

SCHEDULE SE (Form 1040)

Department of the Treasury Internal Revenue Service (0)

Self-Employment Tax

▶ See Instructions for Schedule SE (Form 1040).

▶ Attach to Form 1040.

OMB No. 1545-0074

1992

Attachment Sequence No. **17**

Name of person with **self-employment** income (as shown on Form 1040)	Social security number of person with **self-employment** income ▶	

Who Must File Schedule SE

You must file Schedule SE if:

- Your wages (and tips) subject to social security AND Medicare tax (or railroad retirement tax) were less than $130,200; **AND**
- Your *net earnings from self-employment from other than church employee income* (line 4 of Short Schedule SE or line 4c of Long Schedule SE) were $400 or more;

OR

- You had church employee income (as defined on page SE-1) of $108.28 or more.

Exception. If your only self-employment income was from earnings as a minister, member of a religious order, or Christian Science practitioner, AND you filed **Form 4361** and received IRS approval not to be taxed on those earnings, DO NOT file Schedule SE. Instead, write "Exempt–Form 4361" on Form 1040, line 47.

May I Use Short Schedule SE or MUST I Use Long Schedule SE?

Did you receive wages or tips in 1992?

No → Are you a minister, member of a religious order, or Christian Science practitioner who received IRS approval **not** to be taxed on earnings from these sources, **but** you owe self-employment tax on other earnings?
- Yes → YOU MUST USE LONG SCHEDULE SE ON THE BACK
- No → Are you using one of the optional methods to figure your net earnings (see page SE-3)?
  - Yes → YOU MUST USE LONG SCHEDULE SE ON THE BACK
  - No → Did you receive church employee income reported on Form W-2 of $108.28 or more?
    - Yes → YOU MUST USE LONG SCHEDULE SE ON THE BACK
    - No → **YOU MAY USE SHORT SCHEDULE SE BELOW**

Yes → Was the total of your wages and tips subject to social security or railroad retirement tax **plus** your net earnings from self-employment more than $55,500?
- Yes → YOU MUST USE LONG SCHEDULE SE ON THE BACK
- No → Was the total of your wages and tips subject to Medicare tax **plus** your net earnings from self-employment more than $130,200?
  - Yes → YOU MUST USE LONG SCHEDULE SE ON THE BACK
  - No → Did you receive tips subject to social security or Medicare tax that you **did not** report to your employer?
    - Yes → YOU MUST USE LONG SCHEDULE SE ON THE BACK
    - No → Are you a minister, member of a religious order, or Christian Science practitioner…? (continue with left column)

YOU MUST USE LONG SCHEDULE SE ON THE BACK

Section A—Short Schedule SE. Caution: *Read above to see if you must use Long Schedule SE on the back (Section B).*

1	Net farm profit or (loss) from Schedule F, line 36, and farm partnerships, Schedule K-1 (Form 1065), line 15a	1		
2	Net profit or (loss) from Schedule C, line 31; Schedule C-EZ, line 3; and Schedule K-1 (Form 1065), line 15a (other than farming). See page SE-2 for other income to report	2		
3	Combine lines 1 and 2	3		
4	**Net earnings from self-employment.** Multiply line 3 by 92.35% (.9235). If less than $400, **do not** file this schedule; you do not owe self-employment tax ▶	4		
5	**Self-employment tax.** If the amount on line 4 is: • $55,500 or less, multiply line 4 by 15.3% (.153) and enter the result. • More than $55,500 but less than $130,200, multiply the amount in excess of $55,500 by 2.9% (.029). Then, add $8,491.50 to the result and enter the total. • $130,200 or more, enter $10,657.80. Also, enter this amount on Form 1040, line 47	5		

Note: *Also, enter **one-half** of the amount from line 5 on **Form 1040, line 25**.*

For Paperwork Reduction Act Notice, see Form 1040 instructions. Cat. No. 11358Z **Schedule SE (Form 1040) 1992**

A-11a Form 1040: Self-Employment Tax (front).

Department of the Treasury—Internal Revenue Service

Schedule SE (Form 1040) 1992 — Attachment Sequence No. **17** — Page **2**

Name of person with **self-employment** income (as shown on Form 1040) | Social security number of person with **self-employment** income ▶

Section B—Long Schedule SE

A If you are a minister, member of a religious order, or Christian Science practitioner AND you filed **Form 4361,** but you had $400 or more of **other** net earnings from self-employment, check here and continue with Part I . . . . . . ▶ ☐

B If your only income subject to self-employment tax is church employee income and you are **not** a minister or a member of a religious order, skip lines 1 through 4b. Enter -0- on line 4c and go to line 5a.

Part I Self-Employment Tax

Line	Description	Box	Amount	Cents
1	Net farm profit or (loss) from Schedule F, line 36, and farm partnerships, Schedule K-1 (Form 1065), line 15a. **Note:** *Skip this line if you use the farm optional method. See requirements in Part II below and on page SE-3*	1		
2	Net profit or (loss) from Schedule C, line 31; Schedule C-EZ, line 3; and Schedule K-1 (Form 1065), line 15a (other than farming). See page SE-2 for other income to report. **Note:** *Skip this line if you use the nonfarm optional method. See requirements in Part II below and on page SE-3*	2		
3	Combine lines 1 and 2	3		
4a	If line 3 is more than zero, multiply line 3 by 92.35% (.9235). Otherwise, enter amount from line 3	4a		
b	If you elected one or both of the optional methods, enter the total of lines 17 and 19 here	4b		
c	Combine lines 4a and 4b. If less than $400, **do not** file this schedule; you do not owe self-employment tax. **Exception.** If less than $400 and you had church employee income, enter -0- and continue ▶	4c		
5a	Enter your church employee income from Form W-2. **Caution:** *See page SE-1 for definition of church employee income* [5a]			
b	Multiply line 5a by 92.35% (.9235). If less than $100, enter -0-	5b		
6	**Net earnings from self-employment.** Add lines 4c and 5b	6		
7	Maximum amount of combined wages and self-employment earnings subject to social security tax or the 6.2% portion of the 7.65% railroad retirement (tier 1) tax for 1992	7	55,500	00
8a	Total social security wages and tips (from Form(s) W-2) and railroad retirement (tier 1) compensation [8a]			
b	Unreported tips subject to social security tax (from Form 4137, line 9) [8b]			
c	Add lines 8a and 8b	8c		
9	Subtract line 8c from line 7. If zero or less, enter -0- here and on line 10 and go to line 12a ▶	9		
10	Multiply the **smaller** of line 6 or line 9 by 12.4% (.124)	10		
11	Maximum amount of combined wages and self-employment earnings subject to Medicare tax or the 1.45% portion of the 7.65% railroad retirement (tier 1) tax for 1992	11	130,200	00
12a	Total Medicare wages and tips (from Form(s) W-2) and railroad retirement (tier 1) compensation [12a]			
b	Unreported tips subject to Medicare tax (from Form 4137, line 14) [12b]			
c	Add lines 12a and 12b	12c		
13	Subtract line 12c from line 11. If zero or less, enter -0- here and on line 14 and go to line 15	13		
14	Multiply the **smaller** of line 6 or line 13 by 2.9% (.029)	14		
15	**Self-employment tax.** Add lines 10 and 14. Enter the result here and on Form 1040, line 47	15		

Note: *Also, enter **one-half** of the amount from line 15 on **Form 1040, line 25.***

Part II Optional Methods To Figure Net Earnings (See **Who Can File Schedule SE** on page SE-1 and **Optional Methods** on page SE-3.)

Farm Optional Method. You may use this method **only** if **(a)** Your gross farm income[1] was not more than $2,400 **or (b)** Your gross farm income[1] was more than $2,400 and your net farm profits[2] were less than $1,733.

Line	Description	Box	Amount	Cents
16	Maximum income for optional methods	16	1,600	00
17	Enter the **smaller** of: two-thirds (⅔) of gross farm income[1] **or** $1,600. Also, include this amount on line 4b above	17		

Nonfarm Optional Method. You may use this method **only** if **(a)** Your net nonfarm profits[3] were less than $1,733 and also less than 72.189% of your gross nonfarm income,[4] **and (b)** You had net earnings from self-employment of at least $400 in 2 of the prior 3 years. **Caution:** *You may use this method no more than five times.*

Line	Description	Box	Amount	Cents
18	Subtract line 17 from line 16	18		
19	Enter the **smaller** of: two-thirds (⅔) of gross nonfarm income[4] **or** the amount on line 18. Also, include this amount on line 4b above	19		

[1]From Schedule F, line 11, and Schedule K-1 (Form 1065), line 15b.
[2]From Schedule F, line 36, and Schedule K-1 (Form 1065), line 15a.
[3]From Schedule C, line 31; Schedule C-EZ, line 3; and Schedule K-1 (Form 1065), line 15a.
[4]From Schedule C, line 7; Schedule C-EZ, line 1; and Schedule K-1 (Form 1065), line 15c.

*U.S. GPO: 1992-315-191

A-11b Form 1040: Schedule SE Self-Employment Tax (back).

Department of the Treasury—Internal Revenue Service

Form **4797** | **Sales of Business Property** **(Also Involuntary Conversions and Recapture Amounts Under Sections 179 and 280F)** ▶ Attach to your tax return. ▶ See separate instructions. | OMB No. 1545-0184 **1991** Attachment Sequence No. **27**

Department of the Treasury Internal Revenue Service (0)

Name(s) shown on return | Identifying number

Part I **Sales or Exchanges of Property Used in a Trade or Business and Involuntary Conversions From Other Than Casualty or Theft—Property Held More Than 1 Year**

1 Enter here the gross proceeds from the sale or exchange of real estate reported to you for 1991 on Form(s) 1099-S (or a substitute statement) that you will be including on line 2, 10, or 20 1

(a) Description of property	(b) Date acquired (mo., day, yr.)	(c) Date sold (mo., day, yr.)	(d) Gross sales price	(e) Depreciation allowed or allowable since acquisition	(f) Cost or other basis, plus improvements and expense of sale	(g) LOSS ((f) minus the sum of (d) and (e))	(h) GAIN ((d) plus (e) minus (f))
2							

3 Gain, if any, from Form 4684, Section B, line 21 . . . .

4 Section 1231 gain from installment sales from Form 6252, line 22 or 30 . . . .

5 Gain, if any, from line 32, from other than casualty or theft . . . .

6 Add lines 2 through 5 in columns (g) and (h) . . . . ( )

7 Combine columns (g) and (h) of line 6. Enter gain or (loss) here, and on the appropriate line as follows: . . . . .

Partnerships.—Enter the gain or (loss) on Form 1065, Schedule K, line 6. Skip lines 8, 9, 11, and 12 below.

S corporations.—Report the gain or (loss) following the instructions for Form 1120S, Schedule K, lines 5 and 6. Skip lines 8, 9, 11, and 12 below, unless line 7 is a gain and the S corporation is subject to the capital gains tax.

All others.—If line 7 is zero or a loss, enter the amount on line 11 below and skip lines 8 and 9. If line 7 is a gain and you did not have any prior year section 1231 losses, or they were recaptured in an earlier year, enter the gain as a long-term capital gain on Schedule D and skip lines 8, 9, and 12 below.

8 Nonrecaptured net section 1231 losses from prior years (see instructions) . . . .

9 Subtract line 8 from line 7. If zero or less, enter -0-. Also enter on the appropriate line as follows (see instructions): .

S corporations.—Enter this amount (if more than zero) on Schedule D (Form 1120S), line 7, and skip lines 11 and 12 below.

All others.—If line 9 is zero, enter the amount from line 7 on line 12 below. If line 9 is more than zero, enter the amount from line 8 on line 12 below, and enter the amount from line 9 as a long-term capital gain on Schedule D.

Part II **Ordinary Gains and Losses**

10 Ordinary gains and losses not included on lines 11 through 16 (include property held 1 year or less):

11 Loss, if any, from line 7 . . . .

12 Gain, if any, from line 7, or amount from line 8 if applicable . . . .

13 Gain, if any, from line 31 . . . .

14 Net gain or (loss) from Form 4684, Section B, lines 13 and 20a . . . .

15 Ordinary gain from installment sales from Form 6252, line 21 or 29 . . . .

16 Recapture of section 179 deduction for partners and S corporation shareholders from property dispositions by partnerships and S corporations (see instructions) . . . .

17 Add lines 10 through 16 in columns (g) and (h) . . . . ( )

18 Combine columns (g) and (h) of line 17. Enter gain or (loss) here, and on the appropriate line as follows: . . . . .

a For all except individual returns: Enter the gain or (loss) from line 18 on the return being filed.

b For individual returns:

(1) If the loss on line 11 includes a loss from Form 4684, Section B, Part II, column (b)(ii), enter that part of the loss here and on line 20 of Schedule A (Form 1040). Identify as from "Form 4797, line 18b(1)." See instructions . . .

(2) Redetermine the gain or (loss) on line 18, excluding the loss, if any, on line 18b(1). Enter here and on Form 1040, line 15 . . .

For Paperwork Reduction Act Notice, see page 1 of separate instructions. Cat. No. 13086I Form **4797** (1991)

A-12a Form 4797: Sales of Business Property (front).

Department of the Treasury—Internal Revenue Service

Form 4797 (1991) Page 2

Part III Gain From Disposition of Property Under Sections 1245, 1250, 1252, 1254, and 1255

19 Description of section 1245, 1250, 1252, 1254, or 1255 property:	Date acquired (mo., day, yr.)	Date sold (mo., day, yr.)
A		
B		
C		
D		

Relate lines 19A through 19D to these columns ▶	Property A	Property B	Property C	Property D
20 Gross sales price (**Note:** *See line 1 before completing.*)				
21 Cost or other basis plus expense of sale				
22 Depreciation (or depletion) allowed or allowable				
23 Adjusted basis. Subtract line 22 from line 21				
24 Total gain. Subtract line 23 from line 20				
25 **If section 1245 property:**				
a Depreciation allowed or allowable from line 22				
b Enter the **smaller** of line 24 or 25a				
26 **If section 1250 property:** If straight line depreciation was used, enter -0- on line 26g unless you are a corporation subject to section 291.				
a Additional depreciation after 1975 (see instructions)				
b Applicable percentage multiplied by the **smaller** of line 24 or line 26a (see instructions)				
c Subtract line 26a from line 24. If line 24 is not more than line 26a, skip lines 26d and 26e				
d Additional depreciation after 1969 and before 1976				
e Applicable percentage multiplied by the **smaller** of line 26c or 26d (see instructions)				
f Section 291 amount (corporations only)				
g Add lines 26b, 26e, and 26f				
27 **If section 1252 property:** Skip this section if you did not dispose of farmland or if you are a partnership.				
a Soil, water, and land clearing expenses				
b Line 27a multiplied by applicable percentage (see instructions)				
c Enter the **smaller** of line 24 or 27b				
28 **If section 1254 property:**				
a Intangible drilling and development costs, expenditures for development of mines and other natural deposits, and mining exploration costs (see instructions)				
b Enter the **smaller** of line 24 or 28a				
29 **If section 1255 property:**				
a Applicable percentage of payments excluded from income under section 126 (see instructions)				
b Enter the **smaller** of line 24 or 29a				

Summary of Part III Gains (Complete property columns A through D, through line 29b before going to line 30.)

30 Total gains for all properties. Add columns A through D, line 24	
31 Add columns A through D, lines 25b, 26g, 27c, 28b, and 29b. Enter here and on line 13. (See the instructions for Part IV if this is an installment sale.)	
32 Subtract line 31 from line 30. Enter the portion from casualty or theft on Form 4684, Section B, line 15. Enter the portion from other than casualty or theft on Form 4797, line 5	

Part IV Election Not to Use the Installment Method (Complete this part only if you elect out of the installment method and report a note or other obligation at less than full face value.)

33 Check here if you elect out of the installment method ▶ ☐

34 Enter the face amount of the note or other obligation ▶ $

35 Enter the percentage of valuation of the note or other obligation ▶ %

Part V Recapture Amounts Under Sections 179 and 280F When Business Use Drops to 50% or Less (See instructions for Part V.)

	(a) Section 179	(b) Section 280F
36 Section 179 expense deduction or depreciation allowable in prior years		
37 Recomputed depreciation (see instructions)		
38 Recapture amount. Subtract line 37 from line 36. (See instructions for where to report.)		

*U.S. Government Printing Office: 1991 — 285-341

A-12b Form 4797: Sales of Business Property (back).

Department of the Treasury—Internal Revenue Service

Index

D

N

O

P

U

V

W

Y